Lecture Notes in Statistics 161

Edited by P. Bickel, P. Diggle, S. Fienberg, K. Krickeberg,
I. Olkin, N. Wermuth, S. Zeger

T0138084

Springer
New York
Berlin
Heidelberg
Barcelona
Hong Kong
London
Milan
Paris
Singapore
Tokyo

Maarten Jansen

Noise Reduction by Wavelet Thresholding

 Springer

Maarten Jansen
Katholieke Universiteit Leuven
Departement Computerwetenschappen
Celestijnenlaan 200A
B-3001 Heverlee
Belgium
maarten.jansen@cs.kuleuven.ac.be
www.cs.kuleuven.ac.be/~maarten/

Library of Congress Cataloging-in-Publication Data
Jansen, Maarten.
    Noise reduction by wavelet thresholding / Maarten Jansen.
        p. cm. — (Lecture notes in statistics ; 161)
    Includes bibliographical references and index.
    ISBN 0-387-95244-6 (softcover : alk. paper)
    1. Signal processing— Digital techniques—Statistical methods. 2. Electronic
    noise—Automatic control. 3. Wavelets (Mathematics).   I. Title.   II. Lecture notes
    in statistics (Springer-Verlag) ; v. 161.
TK5102.9.J36 2001
621.382'2—dc21   00-069241

Printed on acid-free paper.

Camera-ready copy provided by the author.
Printed and bound by Sheridan Books, Inc., Ann Arbor, MI.
Printed in the United States of America.

9 8 7 6 5 4 3 2 1

ISBN 0-387-95244-6   SPIN 10793867

Springer-Verlag   New York Berlin Heidelberg
A member of BertelsmannSpringer Science+Business Media GmbH

Preface

Wavelet methods have become a widely spread tool in signal and image processing tasks. This book deals with statistical applications, especially wavelet based
smoothing. The methods described in this text are examples of non-linear and non-
parametric curve fitting. The book aims to contribute to the field both among statisticians and in the application oriented world (including but not limited to signals and
images). Although it also contains extensive analyses of some existing methods, it
has no intention whatsoever to be a complete overview of the field: the text would
show too much bias towards my own algorithms. I rather present new material and
own insights in the questions involved with wavelet based noise reduction. On the
other hand, the presented material does cover a whole range of methodologies, and
in that sense, the book may serve as an introduction into the domain of wavelet
smoothing.

Throughout the text, three main properties show up ever again: sparsity, locality
and multiresolution. Nearly all wavelet based methods exploit at least one of these
properties in some or the other way.

These notes present research results of the Belgian Programme on Interuniver-
sity Poles of Attraction, initiated by the Belgian State, Prime Minister's Office for
Science, Technology and Culture. The scientific responsibility rests with me. My
research was financed by a grant (1995 – 1999) from the Flemish Institute for the
Promotion of Scientific and Technological Research in the Industry (IWT). Since
October 2000, I have been postdoctoral research fellow of the Fund of Scientific
Research Flanders (Belgium) (FWO).

Most of the figures were constructed with the aid of Matlab^{TM} and the software package WaveLab, developed at Stanford University, department of Statistics.
I gradually developed my own matlab routines for wavelet transforms and noise reduction. This collection is now available on the world wide web, and this suffices to
reproduce the Matlab figures throughout the chapters. More information comes at
the end of this volume. I also used the software library WAILI for integer wavelet
transforms, developed at K.U.Leuven, Department of Computer Science [168].

I would like to thank all wavelet colleagues at K.U.Leuven Leuven and Bristol
University for their support and discussions.

Leuven, Belgium   Maarten Jansen
November, 2000

Contents

Notations and Abbreviations

List of Symbols

| ω | : | wavelet coefficients of input noise | 36 |
| ω_λ | : | wavelet coefficients of output noise | 48 |

List of Abbreviations

1D	:	one-dimensional	24
2D	:	two-dimensional	36
AWGN	:	Additive, white, Gaussian noise	35
CDF	:	Cohen-Daubechies-Feauveau (wavelets)	22
CPRESS	:	Complexity-penalized residual sum of squares	110
CT	:	Computed Tomography	7
CWT	:	Continuous wavelet transform	25
DSA	:	Digital Subtraction Angiography	129
DTFT	:	Discrete Time Fourier Transform	5
DWT	:	Discrete Wavelet Transform	25
EM	:	Expectation-Minimization	157
FDR	:	False Discovery Rate	63
FIR	:	Finite Impulse Response (filter)	18
FFT	:	Fast Fourier Transform	5
FWT	:	Fast Wavelet Transform	19
GCV	:	Generalized cross validation	87
GIS	:	Geographical information system	6
HVS	:	Human Visual System	7
MAD	:	Median Absolute Deviation	124
MAP	:	Maximum A Posteriori	153
MCMC	:	Markov Chain Monte Carlo	154
MDL	:	Minimum Description Length	38
MLE	:	Maximum Likelihood Estimation	156
MMP	:	Maximal Marginal Posterior	153
MPLE	:	Maximum Pseudo-Likelihood Estimation	156
MRA	:	Multiresolution analysis	13
MRF	:	Markov Random Field	145
MRI	:	Magnetic Resonance Imaging	7
MSE	:	Mean Square Error	48
PSNR	:	Peak signal-to-noise ratio	48
SNR	:	Signal-to-noise ratio	48
SURE	:	Stein's Unbiased Risk Estimate	82
a.s.	:	almost sure(ly) (with probability one)	94
i.i.d.	:	independent, identically distributed noise	35
dB	:	deciBel	48

List of Figures

List of Tables

1

Introduction and overview

Thanks to the combination of a nice theoretical foundation and the promising applications, wavelets have become a popular tool in many research domains. In fact, wavelet theory combines many existing concepts into a global framework. This new theoretical basis reveals new insights and throws a new light on several domains of applications.

This text is situated on the bridge between two or three domains of application: on one side, we have statistics, on the other side there are the domains of digital signal- and image processing. From time to time, we encounter notions from approximation theory too.

1.1 Outline and situation

This text is about noise reduction or non-parametric regression, using wavelets. It focuses on the technique of wavelet thresholding. This method is simple and fast. Chapter 2 explains the essentials about wavelets and wavelet based noise reduction. Three main wavelet properties play an important rule in smoothing algorithms: sparsity, locality and multiresolution. Only the last corresponds to strict mathematical concept. The first two rather appear as assumptions for proofs of theorems or algorithms: typically, results are only valid under certain conditions, and these conditions can then be explained as sparsity and/or locality. Chapter 2 also introduces the idea of wavelet thresholding and addresses the problems involved with this method.

The next two chapters belong together. Chapter 3 studies the asymptotic behavior of the threshold that minimizes the expected mean square error (MSE) of the result.It proves that for piecewise polynomials, this threshold grows asymptotically as:

$$\lambda^* \sim \sqrt{2 \log N} \sigma,$$

if the number of samples N tends to infinity. This is exactly the same expression as for the *universal threshold*. For piecewise Lipschitz α functions, this needs a slight correction:

$$\lambda^* \sim \sqrt{\frac{2\alpha}{2\alpha + 1}} \sqrt{2 \log N} \sigma.$$

The mean square error is of course not the only possible objective function for a noise reduction algorithm nor does it yield the best output in all circumstances. In spite of its shortcomings, it is often used, because of its mathematical tractability and fair results in a wide range of applications.

Since the noise-free data are unknown in practical situations, we cannot compute and minimize the mean square error of the output exactly. Therefore, Chapter 4 presents an estimation of this error function, based on the method of generalized cross validation. This method is well known in the framework of linear regression, like spline smoothing [174]. Weyrich and Warhola [177] formulated the definition of GCV for wavelet-thresholding applied to noisy observations y:

$$GCV(\lambda) = \frac{\frac{1}{N} \cdot \|y - y_\lambda\|^2}{\left(\frac{N_0(\lambda)}{N}\right)^2}.$$

$GCV(\lambda)$ is a function of the threshold value λ through the output y_λ and the number of killed coefficients $N_0(\lambda)$. We use the asymptotic results from Chapter 3 to prove minimizing this GCV is an asymptotcally optimal procedure. More precisely, Chapter 4 shows that if $\lambda^* = arg \min MSE(\lambda)$ and $\hat{\lambda} = arg \min GCV(\lambda)$, then for $N \to \infty$, both minimizers yield a result of the same quality:

$$\frac{\mathrm{E} MSE(\hat{\lambda})}{\mathrm{E} MSE(\lambda^*)} \to 1.$$

The proof is inspired by the spline proof by Wahba [174], but the non-linear character of the threshold operation caused several additional problems. We also show that GCV minimisation is faster than the fast wavelet transform (except for the Haar transform).

The proof of the asymptotic optimality of GCV not only motivates its use in the standard setting of white, stationary noise and orthogonal transforms. After two chapters of motivation, we investigate the applicability of the method in less conventional situations. This includes colored (correlated) noise, non-orthogonal transforms, non-decimated transforms. We argue that the advantages of GCV may play an important role, for instance in level-dependent thresholding, to avoid an explicit noise variance estimation at each level. We also consider more general coefficient selection procedures, other than simple thresholding on coefficient magnitudes. Especially, we pay attention to interscale correlations and tree structured thresholding. Whereas Chapters 3 and 4 mainly rely on the *sparsity* of a wavelet transform, Chapter 5 also uses the concept of *multiresolution*, naturally supported by wavelet theory. This more practice-oriented chapter also investigates noise reduction for integer transforms, which are interesting for fast and lossless algorithms in applications like image processing.

Chapter 6 concentrates on two-dimensional problems, such as *image* noise reduction. We illustrate how additional difficulties show up when proceeding from one dimension to two dimensions: line singularities (e.g. edges) do not exist in one dimension and create intrascale correlations. We propose a Bayesian approach based on a geometrical prior for configurations of important wavelet coefficients. This prior is typically a Markov Random Field (MRF). By incorporating geometrical information, we hope to mimic as well as possible the optimal coefficient selection.

The last chapter is dedicated to regression of non-equispaced data. Our approach is new in the sense that it examines the applicability of so-called second generation wavelets in the context of noise reduction. This extension of classical wavelets was developed with applications on irregular grids in mind. When using this second generation transforms, we encounter several stability problems, and although these troubles do not show up in cases with "not too irregular" samples, we carefully design an algorithm combining stability of classical wavelet transforms and smoothness of the new transforms.

1.2 Notions and notations

Before actually describing wavelets and their applications in the next chapter, we first introduce some concepts and notations from the various fields. This chapter is merely technical and maybe skipped for first reading. If concepts in the subsequent chapters are unclear, there is a good chance that a definition or description can be looked up in this section.

1.2.1 Mathematical preliminaries

A real, digital signal f is nothing but a sequence of real numbers: $f_i \in \mathbb{R}$. In this text, we suppose them to be square summable: $f \in \ell_2(\mathbb{Z})$, where $\mathbb{Z}$ denotes the set of integers. In most practical applications, the signals have a finite number N of non-zero elements. With a slight abuse of notation, we could say that $f \in \mathbb{R}^N \subset \ell_2(\mathbb{Z})$, $\mathbb{R}$ being the set of real numbers.

In a way that we explain immediately, such a discrete input can be associated with a function defined on the interval $[0, 1]$. This function is square integrable, *das heißt*:

$$f(x) \in L_2[0, 1] \Leftrightarrow \int_{[0,1]} f^2(x)dx < \infty.$$

Strictly speaking, this integral should be taken in the sense of *Lebesgue*, although the Riemann construction suffices to understand what follows. The space of all square integrable functions is a Hilbert space. This means that it is a unitary space, *id est* a complete vector space with a definition of a *scalar product* (inner product or dot product):

$$\langle f, g \rangle = \int_{[0,1]} f(x)g(x)dx.$$

Such an inner product allows for the notion of orthogonality: two elements (functions) are said to be orthogonal if their inner product equals zero. An inner product also induces a *norm*:

$$\|f\| = \sqrt{\langle f, f \rangle}.$$

A Hilbert space must also be complete, *cioè*, all Cauchy sequences must converge within the Hilbert space and with respect to its norm. A Cauchy sequence is a sequence of functions which come arbitrarily close to each other, with respect to the given norm.

Square integrability can be generalized:

$$f(x) \in L_p[0,1] \Leftrightarrow \int_{[0,1]} |f(x)|^p dx < \infty,$$

but for $p \neq 2$, there is no scalar product which induces the norm

$$\|f\|_p = \int_{[0,1]} |f(x)|^p dx.$$

The function spaces $L_p[0,1]$ with $p \geq 1$ are examples of Banach spaces, complete vector spaces with a norm, but not necessarily an inner product.

A *basis* $\varphi_k, k \in \mathbb{Z}$ is a free set of functions (*c'est-à-dire* a set of linearly independent functions) that generates the entire space. Since we are dealing with infinite dimensional spaces, we should we careful about the word *generate*: we are dealing with infinite sums and convergence issues are involved [59]: if there exists for every function $f \in L_p[0,1]$ a unique sequence $s_i, i \in \mathbb{Z}$ so that $f(x) = \sum_i s_i \varphi_i(x)$, then we have a *Schauder basis*. This uniqueness is the guarantee for linear independence of the basis functions, but convergence in such a basis may depend on the ordering of the components. A basis is called *unconditional* if $\sum_i |s_i| \varphi_i(x) \in L_p[0,1]$ for all $\sum_i s_i \varphi_i(x) \in L_p[0,1]$ and *vice versa*. As a consequence, the sum $\sum_i s_i \varphi_i(x)$ converges independent of the order of summation. Unconditional basis means that the coefficient *magnitude* only determines whether or not a function belongs to a Banach space: the phase of the coefficients (in real analysis this is the sign only) is of no importance.

In the Hilbert space $L_2[0,1]$, an unconditional basis is called a *Riesz* basis if it is *almost normalized*. This means that there exist real, positive, non-zero constants m and M so that:

$$0 < m \leq \|\varphi_i\| \leq M < \infty.$$

A Riesz basis is characterized by two Riesz *constants* A and B, so that for all $f = \sum_i s_i \varphi_i \in L_2[0,1]$:

$$A^2\|f\|^2 \leq \sum_{i \in \mathbb{Z}} s_i^2 \leq B^2\|f\|^2.$$

A Riesz basis is also called a *stable* basis. It is essentially the second best type of basis after orthonormal bases.

1.2.2 Fourier analysis and digital signals

Functions in $L_2[0, 1]$ can be decomposed into a basis of waves $e^{i2\pi kx}$:

$$f(x) = \sum_{k \in \mathbb{Z}} c_k e^{i2\pi kx}.$$

This is a Fourier series expansion. Since these waves constitute an orthogonal basis, the coefficients are easy to find:

$$c_k = \frac{1}{2\pi} \int_0^1 f(x)e^{-i2\pi kx}.$$

The minus sign in the exponent appears because the basis functions are complex, and the proper definition of a scalar product for complex functions uses complex conjugates. The basis functions are of course not limited to the interval $[0, 1]$: they are periodic. The Fourier series is also valid for a periodic extension of $f \in L_2[0, 1]$ to the entire real axis.

General functions in $L_2(\mathbb{R})$ are not periodic nor can they be periodically extended. Frequency analysis now goes by a *Fourier transform*, defined as:

$$F(\omega) = \mathcal{F}\{f(x)\} = \frac{1}{\sqrt{2\pi}} \int_{-\infty}^{\infty} f(x)e^{-i\omega x}dx.$$

The inverse of this transform is given by:

$$f(x) = \frac{1}{\sqrt{2\pi}} \int_{-\infty}^{\infty} F(\omega)e^{i\omega x}d\omega.$$

Since most of this text is about discrete signals (samples), we are also investigating the frequency contents of this kind of signals. This is given by the inner product of a discrete signal f with a wave $\{e^{i\omega k}\}$:

$$F(\omega) = \sum_{k \in \mathbb{Z}} f_k e^{-i\omega k}.$$

This is the Discrete Time Fourier Transform (DTFT). By a substitution $\omega = -2\pi x$, we see that f_k is nothing but the Fourier series expansion of $F(\omega)$. $F(\omega)$ is 2π-periodic and the inverse of a DTFT is a Fourier series expansion. Discretizing the frequency parameter ω in a DTFT leads to a (fully) discrete Fourier transform (DFT), for which exists a fast algorithm, the Fast Fourier Transform (FFT) .

A DTFT gives the frequency *contents* of a discrete signal, but, as the next chapter illustrates, it is also the formula for the frequency *response* of a linear, time-invariant filter. A digital *filter* $\mathcal{H}$ is any system operating on a digital signal. Linear filters satisfy:

$$\mathcal{H}(\alpha f + \beta g) = \alpha \mathcal{H}(f) + \beta \mathcal{H}(g)$$

and time-invariance (or shift-invariance) means that: $\mathcal{H}D = D\mathcal{H}$, where D is the delay (inverse shift) operator:

$$y = Dx \Leftrightarrow y_k = x_{k-1}, \forall k \in \mathbb{Z}.$$

A time-invariant filter is characterized by its *impulse response* h:

$$h = \mathcal{H}(\delta_0),$$

where $\delta_0 = \{\ldots, 0, 0, 1, 0, 0, \ldots\}$ is a Kronecker delta (discrete Dirac impulse) signal. h by itself is a signal. If the filter is linear, its response to an arbitrary signal x equals:

$$y = \mathcal{H}\left(\sum_{k=-\infty}^{\infty} x_k \delta_k\right) = \sum_{k=-\infty}^{\infty} x_k \mathcal{H}(D^k \delta_0) = \sum_{k=-\infty}^{\infty} x_k D^k h.$$

and so:

$$y_n = \sum_{k=-\infty}^{\infty} h_{n-k} x_k.$$

This expression is called a *convolution sum*.

The next chapter illustrates with an example how a filter modifies the frequency contents of a signal.

1.2.3 A note on images

A digital image can be seen as a matrix of pixels. Digital image *processing* is everything *beyond* a matrix of pixels. It deals with all the operations you can perform on the image by considering the image *not* just as a matrix. To distinguish an image, which is basically a specially structured data *vector* from a linear operation matrix, we use bold lower case letters to denote an image, except when this image is a random variable.

So x is the normal notation for an image, X emphasizes that this image is a matrix of random variables and X may denote a random variable or a linear operation matrix.

Images play an important role, both in daily life applications and in areas of research and technology. We mention geographical information systems (GIS), astronomical and medical images. It is true that image acquisition techniques, like cameras, microscopes and various types of scanners (CT - computed tomography, MRI - magnetic resonance imaging) have had important developments in the last years and that images carry much less noise than before. On the other hand, the requisites for the desired applications are often stronger too. Nowadays, a medical image may have 2048 grey levels instead of the classical 256. This is more than what our human visual system (HVS) can distinguish. Typically, only a small part of this dynamical range is important. For the application, contrast in this part is enhanced so that this interesting piece covers the entire range from black to white. All other grey levels in the original image are mapped to perfect white or black. This contrast enhancement blows up the present noise. So, even if in the first instance the noise is quasi invisible, it may be important to reduce it, in the light of the further use of these data.

1.2.4 Generality

Imaging is an important application of the noise reduction algorithms described in this text. Nevertheless, we emphasize that nearly all the material in this text has a wider range of applications. Since this research was not performed in an image processing laboratory, but in a division of applied mathematics, we chose not to concentrate on this specific domain of application, but rather formulate more general algorithms. The method of Chapter 6 is more image-oriented, but yet, this may serve for other two-dimensionally structured data too. We want too stress this generality also on another point: the procedure of generalized cross validation for non-linear smoothing, as described in Chapter 4 is not limited to wavelets. It should also work for other types of sparse data representations. We discuss in Chapter 6 why wavelets might not be the ultimate way for representing images, but generalized cross validation remains interesting, even beyond the classical wavelets.

Section 2.10 contains a more complete list of wavelet applications. These are not limited to the problem of noise reduction.

2

Wavelets and wavelet thresholding

Every theory starts from an idea. The wavelet idea is simple and clear. At a first confrontation, the mathematics that work out this idea might appear strange and difficult. Nevertheless, after a while, this theory leads to insight in the mechanism in wavelet based algorithms in a variety of applications.

This chapter discusses the wavelet idea and explains the wavelet slogans. For the mathematics, we refer to the numerous publications. Comprehensive introductions to the field include [158, 32, 26, 141]. Other, sometimes more application oriented or more theoretical treatments can be found in [130, 59, 164, 120, 149, 110, 6, 107]. Books on wavelets in statistics, such as [144, 172], include large parts on non-parametric regression.

2.1 Exploiting sample correlations

2.1.1 The input problem: sparsity

Suppose we are given a *discrete* signal f. In practice, this signal is often digital, i.e. quantized and possibly transformed into a binary form. Figure 2.1 shows how these discrete data can be represented on a continuous line as a piecewise constant function. This piecewise constant is of course not the only possible continuous representation.

Typically, adjacent points show strong correlations. Only at a few points, we find large jumps. Storing all these values separately seems a waste of storage capacity. Therefore, we take a pair of neighbors f_1 and f_2 and compute average and difference coefficients:

$$a_1 = \frac{f_1 + f_2}{2}$$

$$d_1 = \frac{f_1 - f_2}{2}$$

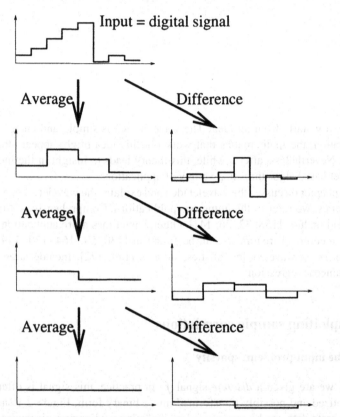

Figure 2.1. Using correlations between neighboring samples leads to a sparse representation of the input. This is the Haar wavelet transform.

In the figure, the averages are represented on the second line as a piecewise constant, just like the input, but the difference coefficients appear as two opposite blocks: every pair of two opposite blocks is one coefficient. This coefficient tells how far the first data point was under the average of the pair and at the same time, how much the second data point was above this average. 'Adding' the left plot and the right one returns the input on top. This 'adding' is indeed the inverse operation:

$$f_1 = a_1 + d_1$$
$$f_2 = a_1 - d_1$$

The average signal is somehow a blurred version of the input. We can repeat the same procedure on the averages again. Eventually, this operation decomposes the input into one global average *plus* difference signal at several locations on the axis and with different widths, *scales* , or *resolutions*. Since each step is invertible, the whole transform satisfies the *perfect reconstruction* property. This is called the *Haar*-transform, after Alfred Haar, who was the first to study it in 1910, long before the actual wavelet history began [90].

As the Figure 2.1 illustrates, most of the difference coefficients are small. The largest coefficient appears at the location of the biggest 'jump' in the input signal. This is even more striking in the more realistic example of Figure 2.2. In this picture, all coefficients are plotted on one line. Dashed lines indicate the boundaries between scales. Only a few coefficients are significant. They indicate the singularities (jumps) in the input. This *sparsity* is a common characteristic for all wavelet transforms. Wavelet transforms are said to have a *decorrelating property*.

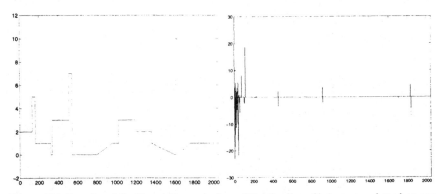

Figure 2.2. Test signal (Left) and Haar transform (Right): all coefficients are plotted on one line. Dashed lines indicate the boundaries between scales.

2.1.2 Basis functions and multiresolution

The input vector f can be seen as coefficients for a basis of characteristic functions ('block' functions), as shown on top of Figure 2.3: i.e. we can write the continuous representation $f(x)$ of the discrete input f as a linear combination of these basis functions, which we call $\varphi_k(x)$:

$$f(x) = \sum_k f_k \varphi_k(x).$$

All these functions are translates of one *father* function, they are called *scaling* functions. The differences, computed during the algorithm, correspond to the basis functions on the next rows in Figure 2.3. All these functions are translations (shifts) *and* dilations (stretches) of one *mother* function. Because these functions have block-wave-form, vanishing outside a small interval, they are called 'short waves' or wave-lets.

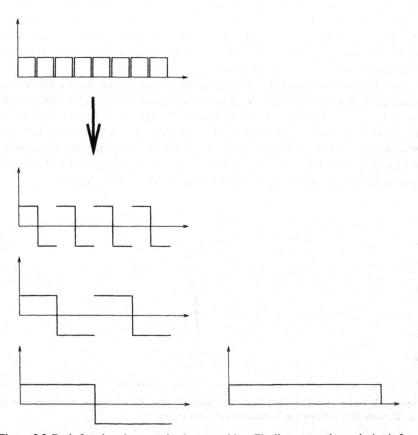

Figure 2.3. Basis functions in a wavelet decomposition. The line on top shows the basis functions of the original representation: any linear combination of these characteristic functions leads to a piecewise constant. Piecewise constants can also be built in a block-wavelet-basis: these basis functions have a short waveform, all have a certain scale and are situated at different locations.

The decomposition from a function in a scaling basis into a wavelet basis is an example of a *multiresolution* analysis. In image processing, the scaling function basis corresponds roughly to the classical pixel representation. Not only is this redundant, our visual system does not look at images that way. The wavelet representation, i.e. a set of details at different locations and different scales, is said to be closer to the way we look at images: at a first sight, we see general features, at a closer inspection, we discover the details.

Instead of just taking averages and differences, one could of course think of more complex tricks to exploit inter-sample coherence. This corresponds to less trivial 'continuous representations of discrete data' than just blocks and blocky wave-lets: this is the way we can build more complex wavelets. While doing so, we have to take care not to lose basic properties, like transform invertibility, and several convergence issues. To this end, we start with a formal definition of the notion of multiresolution:

Definition 2.1 *A sequence of nested, closed subspaces* $V_j \subset L_2[0,1]$, $j = L, \ldots, J$ *is called a* multiresolution analysis *(MRA) if*

$$\forall j \in \mathbb{Z} : V_j \subset V_{j+1}, \tag{2.1}$$

$$\overline{\lim_{j \to \infty} V_j} = \bigcup_{j \in \mathbb{Z}} V_j = L_2[0,1], \tag{2.2}$$

$$\lim_{j \to -\infty} V_j = \bigcap_{j \in \mathbb{Z}} V_j = \{0\}, \tag{2.3}$$

$$f(x) \in V_j \Leftrightarrow f(2x) \in V_{j+1}, j \in \mathbb{Z}, \quad \text{(scale invariance)} \tag{2.4}$$

$$f(x) \in V_0 \Leftrightarrow f(x+k) \in V_0, k \in \mathbb{Z}, \quad \text{(shift invariance)} \tag{2.5}$$

$$\exists \varphi(x) \in V_0 : \{\varphi(x-k)\}_{k \in \mathbb{Z}} \quad \text{is a stable basis for } V_0 \tag{2.6}$$

A multiresolution analysis is a nested set of function spaces. Projecting a given function into these spaces is like using a telescope: every time we proceed to a finer scale, we add more details to the approximation. The fourth and fifth condition states that we look with the same eyes in all directions and at all scales. The last condition about stability shows up because we are working in an infinite dimensional vector space. When working with finite dimensions, all bases are stable bases. In an infinite dimensional space, basis vectors can come arbitrarily close to each other without really coinciding. This is not possible if we impose the condition of a Riesz basis as in Section 1.2.1. For a more extended treatment of this concept and its consequences, we refer to the literature, e.g.[59].

The function $\varphi(x)$ plays the role of father function: all basis functions of V_0 are shifted versions of this function. Trivialiter, $2^{j/2}\varphi(2^j x - k)$ then is a (normalized) basis for V_j. V_1 contains functions that are not included in V_0. To generate all elements in the finer scale V_1, starting from the basis of the coarser scale V_0, we need additional basis functions. These basis functions generate a space W_0 of detail functions. W_0 is not unique: it maybe the orthogonal or an oblique complement, but it holds that all functions in V_1 can be decomposed in the union of the basis of V_0 and the basis of W_0:

$$V_{j+1} = V_j \oplus W_j.$$

This complementary space W_j has a similar structure: it is generated by translations of one mother function, stretched to meet the current scale j. This mother function is known as wavelet function and is often noted as ψ. The following theorem guarantees that if we start with an orthogonal basis, we are able to transform the data into another orthogonal basis:

Theorem 2.1 *[59, 83] If $\{\varphi(x-k)\}_{k\in\mathbb{Z}}$ constitute an orthogonal basis for V_0, then there exists one function $\psi(x)$ such that $\{\psi(x-k)\}_{k\in\mathbb{Z}}$ forms an orthogonal basis for the orthogonal complement W_0 of V_0 in V_1. Furthermore, $\{\psi(2^j x - k)\}_{k\in\mathbb{Z}}$ constitutes an orthogonal basis for the orthogonal complement W_j of V_j in V_{j+1}. And we have that:*

$$\forall L \in \mathbb{Z} : V_L \oplus \overline{\left(\bigoplus_{j=L}^{\infty} W_j\right)} = L_2[0,1]. \qquad (2.7)$$

A general MRA (multiresolution analysis) has no orthogonality. In this case, we look for a *dual* father function $\tilde{\varphi}(x)$, so that:

$$\langle \tilde{\varphi}(x), \varphi(x-l) \rangle = \delta_l.$$

This pair of bases is called *biorthogonal*. Figure 2.4 illustrates this notion in $\mathbb{R}^3$. The coefficients f in the expansion

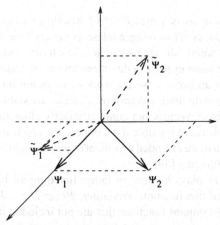

Figure 2.4. A pair of biorthogonal bases in $\mathbb{R}^3$.

$$f(x) = \sum_{k \in \mathbb{Z}} f_k \varphi(x - k)$$

are then:

$$f_k = \langle f(x), \tilde{\varphi}(x - k) \rangle$$

This expression shows that projection in a biorthogonal setting is still simple and stable.

The dual father function generates a dual MRA $\tilde{V}_j$, $j \in \mathbb{Z}$ and the primal wavelet function now also has its dual $\tilde{\psi}(x) \in \tilde{W}_0$, so that:

$$V_j \perp \tilde{W}_j \quad \text{and} \quad \tilde{V}_j \perp W_j,$$

which implies:

$$W_j \perp \tilde{W}_i \quad \text{for } i \neq j$$

This also implies biorthogonality of the basis functions:

$$\langle \tilde{\varphi}(x), \psi(x - k) \rangle = 0 \text{ and } \langle \tilde{\psi}(x), \varphi(x - k) \rangle = 0,$$

and we also have

$$\langle \tilde{\psi}(x), \psi(2^j x - k) \rangle = \delta_j \delta_k.$$

2.1.3 The dilation equation

Wavelet theory, like most interesting branches in mathematics or physics, has a central equation. It is called the *dilation equation, two-scale equation,* or *refinement equation.* It follows from the fact that $\varphi(x) \in V_0 \subset V_1$, so the father function is a linear combination of the basis in V_1:

$$\exists h \in \ell_2(\mathbb{Z}) : \varphi(x) = \sqrt{2} \sum_{k \in \mathbb{Z}} h_k \varphi(2x - k). \tag{2.8}$$

A similar argument holds for the mother function:

$$\exists g \in \ell_2(\mathbb{Z}) : \psi(x) = \sqrt{2} \sum_{k \in \mathbb{Z}} g_k \varphi(2x - k). \tag{2.9}$$

This is the *wavelet equation* . In the case of a biorthogonal basis, there are of course a dual $\tilde{h}$ and a dual $\tilde{g}$. There is a one-to-one relation between these *filters* and the basis functions. Given the filters, the corresponding basis follows by solving dilation and wavelet equation. Solving techniques appear in books like [158]. Since we want a perfectly invertible transform, we cannot combine any four filters h, g, $\tilde{h}$ and $\tilde{g}$. It turns out that there is a strong relation between h and $\tilde{g}$ and between g and $\tilde{h}$ respectively.

We use the following notations for the normalized basis functions:

$$\varphi_{j,k}(x) = 2^{j/2} \varphi(2^j x - k), \tag{2.10}$$

and similarly for $\tilde{\varphi}_{j,k}(x), \psi_{j,k}(x), \tilde{\psi}_{j,k}(x)$. We assume that the two father and the two mother functions are normalized.

2.1.4 (Fast) Wavelet Transforms and Filter Banks

Suppose we want to decompose a signal in a scaling function basis at a given scale into detail coefficients and scaling coefficients at the next, coarser scale: i.e. we have:

$$f_{j+1}(x) = \sum_{k \in \mathbb{Z}} s_{j+1,k} \varphi_{j+1,k}(x).$$

and we want to decompose this into:

$$f_{j+1}(x) = \sum_{k \in \mathbb{Z}} s_{j,k} \varphi_{j,k}(x) + \sum_{k \in \mathbb{Z}} w_{j,k} \psi_{j,k}(x).$$

Computing $s_{j,k}$ and $w_{j,k}$ from $s_{j+1,k}$ is one step in a *Forward Wavelet Transform*. Clearly

$$
\begin{aligned}
w_{j,k} &= \langle f_{j+1}, \tilde{\psi}_{j,k} \rangle \\
&= \langle f_{j+1}, \sum_{l \in \mathbb{Z}} \tilde{g}_l \tilde{\varphi}_{j+1,2k+l} \rangle \\
&= \sum_{l \in \mathbb{Z}} \tilde{g}_{l-2k} s_{j+1,l}.
\end{aligned}
\tag{2.11}
$$

Similarly

$$s_{j,k} = \sum_{l \in \mathbb{Z}} \tilde{h}_{l-2k} s_{j+1,l}. \tag{2.12}$$

The inverse step is easy to find, if we use primal dilation and wavelet equation:

$$
\begin{aligned}
\sum_{l \in \mathbb{Z}} s_{j+1,l} \varphi_{j+1,l} &= \sum_{k \in \mathbb{Z}} s_{j,k} \varphi_{j,k} + \sum_{k \in \mathbb{Z}} w_{j,k} \psi_{j,k} \\
&= \sum_{k \in \mathbb{Z}} s_{j,k} \sum_{l \in \mathbb{Z}} h_l \varphi_{j+1,2k+l} + \sum_{k \in \mathbb{Z}} w_{j,k} \sum_{l \in \mathbb{Z}} g_l \varphi_{j+1,2k+l} \\
&= \sum_{l \in \mathbb{Z}} \varphi_{j+1,l} \left(\sum_{k \in \mathbb{Z}} h_{l-2k} s_{j,k} + \sum_{k \in \mathbb{Z}} g_{l-2k} w_{j,k} \right),
\end{aligned}
$$

from which:

$$s_{j+1,l} = \sum_{k \in \mathbb{Z}} h_{l-2k} s_{j,k} + \sum_{k \in \mathbb{Z}} g_{l-2k} w_{j,k}. \tag{2.13}$$

Forward and inverse transform can be seen as convolution sums, in which (parts of) the dual respectively primal filters appear. It is not a mere convolution. In the reconstruction formula, we only use half of the filter coefficients for the computation of a scaling coefficient: the sum goes over index k, which appears in the expression as

Decomposition   **Reconstruction**

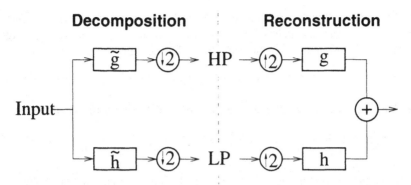

Figure 2.5. One step of a wavelet decomposition and its reconstruction. This is a filter bank: The input is filtered and down-sampled to get a low pass signal *LP* and a high pass signal *HP*. Reconstruction starts with up-sampling by introducing zeroes between every pair of points in *LP* and *HP*.

$2k$: this is up-sampling: we could artificially add zeros between every input scaling or wavelet coefficients and then perform a plain convolution. In the decomposition formula, the sum goes over index l, and k is fixed. This is as if we drop half of the results from a plain convolution. This is *down-sampling*. Putting all this together in a scheme, we get a *filter bank* [170], as in Figure 2.5.

As the symbols HP and LP in the figure indicate, the wavelet filters are typically high pass: they enhance details, whereas the scaling filters are low pass: they have a smoothing effect, eliminate high frequencies. The Haar filters illustrate this: $h = \tilde{h} = \{1/2, 1/2\}$ and $g = \tilde{g} = \{1/2, -1/2\}$. If the input is a pure wave with pulsation ω, i.e.

$$f_k = e^{i\omega k},$$

convolving with h yields an output $y = h * f$, which in general terms equals:

$$y_k = \sum_{l \in \mathbb{Z}} h_l f_{k-l} = \sum_{l \in \mathbb{Z}} h_l e^{i\omega(k-l)} = e^{i\omega k} \sum_{l \in \mathbb{Z}} h_l e^{-i\omega l} = f_k H(\omega).$$

The output is again a pure wave, with the same frequency, but different amplitude and phase. The amplitude depends on the *frequency* ω by (the modulus of) the *frequency response function*

$$H(\omega) = \sum_{l \in \mathbb{Z}} h_l e^{-i\omega l}.$$

This expression is known as the *Discrete Time Fourier Transform* (DTFT). It is actually the inverse of a Fourier series expansion: the function $H(\omega)$ is 2π-periodic. For the Haar scaling function, this becomes:

$$H(\omega) = \frac{1}{2} + \frac{1}{2}e^{-i\omega} = \frac{e^{i\omega/2} + e^{-i\omega/2}}{2} e^{-i\omega/2} = \cos(\frac{\omega}{2})e^{-i\omega/2},$$

for which the amplitude (modulus) $|H(\omega)|$ is plotted on top in Figure 2.6 (Left). This shows that waves with low frequencies ($\omega \approx 0$) are better preserved than high frequencies. Indeed, averaging the strongly oscillating signal

$$\{\ldots, 1, -1, 1, -1, 1, -1, \ldots\},$$

would leave us with a zero output. The situation is different for the detail filter g:

$$G(\omega) = \frac{1}{2} - \frac{1}{2}e^{-i\omega} = \frac{e^{i\omega/2} - e^{-i\omega/2}}{2}e^{-i\omega/2} = \sin(\frac{\omega}{2})ie^{-i\omega/2},$$

and the amplitude plot in Figure 2.6 (Top right) illustrates that low frequencies are being suppressed. A non-oscillating signal, like $\{\ldots, 1, 1, 1, \ldots\}$ has no differences at all: a difference filter shows a zero response to this signal.

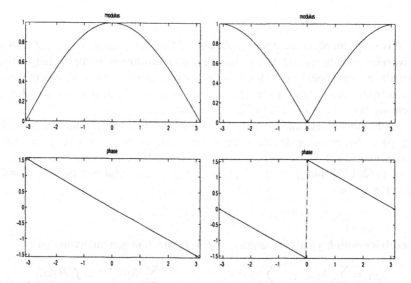

Figure 2.6. Modulus (amplitude) and phase of the frequency response function for the *moving average filter* (Haar low pass filter) $h = \{1/2, 1/2\}$ (Left) and for the *moving difference filter* (Haar high pass filter) $g = \{1/2, -1/2\}$ (Right). Convolution with h suppresses high frequencies, convolution with g suppresses frequencies near to zero.

Figure 2.7 is a schematic overview of a complete wavelet transform of an input signal. In real applications, signals have a finite number N of samples, and also the transform filters have a finite number of taps. Such filters are referred to as *Finite Impulse Response* (FIR) filters. We call $F_{\tilde{h}}$ and $F_{\tilde{g}}$ the length of these filters. From the figure, we conclude that a complete wavelet transform of N data points requires $N-1$ convolutions with $\tilde{h}$ and $N-1$ convolutions with $\tilde{g}$. A convolution with a filter of length F requires F multiplications and $F - 1$ additions. The total complexity of the transform is:

$$(N - 1)(2F_{\tilde{h}} - 1) + (N - 1)(2F_{\tilde{g}} - 1) \sim 2(F_{\tilde{h}} + F_{\tilde{g}})N.$$

A wavelet transform can be computed with a linear amount of flops. Since a general linear transform has square complexity, this is called the *Fast Wavelet Transform* (FWT) [129]. The Fast Fourier Transform (FFT) has complexity $\mathcal{O}(N \log N)$, which is a bit slower than FWT.

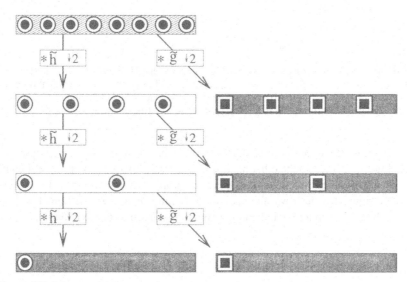

Figure 2.7. Scheme of a Fast Wavelet Transform (FWT). It is computed scale after scale. At each scale, a number of filter operations are needed. Because of subsampling, the number of coefficients to be computed decreases with scale, and this causes the transform to be linear.

2.1.5 Locality

Figure 2.8 illustrates another important feature of a wavelet analysis, which also contributes to its success in applications like noise reduction. Wavelet basis functions are short waves: on one hand, they are well localized in space or time, depending on the actual problem being a signal in time or in space (like images). On the other hand, all basis functions are scaled versions of one mother function. Each basis function therefore corresponds to a unique combination of position and scale. A wavelet coefficient tells how much of the corresponding wavelet basis function 'is present' in the total signal: a high coefficient means that at the given location and scale there is an important contribution of a singularity. This information is local in *space* (or time) and in *frequency* (frequency is approximately the inverse of scale): it says where the singularity (jump in the input) is and how far it ranges, i.e. how large its scale is.

A pixel representation of an image carries no direct scale information: one pixel value gives no information about neighboring pixels, and so there is no notion of

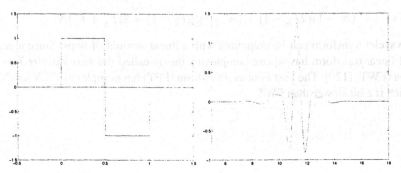

Figure 2.8. Wavelet basis functions for two types of wavelets (Haar and coiflets): these functions live at a specific location and have a specific scale. The coefficient in a signal decomposition that corresponds to this basis function tells how much of this function contributes to the total signal. This information is *local* in space/time *and* frequency.

scale. The basis functions corresponding to this representation are *Dirac* impulses, like in Figure 2.9 (a). On the other hand, a Fourier transform uses pure waves (sines and cosines) as basis functions. It displays a complete frequency spectrum of the image or signal but destroys all space or time information. A coefficient with a never ending wave cannot tell anything about the location of one singularity.

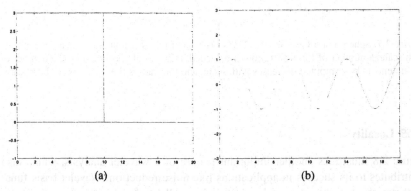

(a)	(b)

Figure 2.9. (a) A Dirac impulse is the basis function behind the classical sample representation of a signal or the pixel representation of an image. One coefficient (sample or pixel) gives the exact information on the function value at this location, but tells nothing about the *range* or *scale* of phenomena happening in the image or signal. (b) A sine has a sharp frequency but is not able to capture any information on the localization of singularities.

No basis function can give exact information on frequency and localization at the same time. Formalizing the notion of *frequency uncertainty* and *space/time uncertainty* leads to lower bound on the product of both. More precisely, define:

$$\overline{x}_\psi = \frac{\int_{-\infty}^{\infty} |\psi(x)|^2 x \, dx}{\int_{-\infty}^{\infty} |\psi(x)|^2 \, dx}$$

$$\Delta x_\psi{}^2 = \frac{\int_{-\infty}^{\infty} |\psi(x)|^2 (x - \overline{x}_\psi)^2 \, dx}{\int_{-\infty}^{\infty} |\psi(x)|^2 \, dx}$$

and let $\overline{\omega}_\Psi$ and $\Delta\omega_\Psi{}^2$ be similar entities in the Fourier domain, then:

$$\Delta x_\psi \Delta\omega_\Psi \geq \frac{1}{2} \text{ or } \Delta x_\psi{}^2 \Delta\omega_\Psi{}^2 \geq \frac{1}{4}.$$

This is Heisenberg's uncertainty principle, mainly known from physics, but actually a purely mathematical inequality [158].

Not only a wavelet coefficient carries local *information*, manipulating it causes a local *effect*, both in space and in frequency. Signal or image processing by operating on wavelet coefficients permits good control on what the algorithm is actually doing.

The idea of locality in time *and* frequency is far from new. The notes of a music score for instance indicate which tone (frequency) should sound at a given moment (time). One could phrase that the score is an approximate wavelet transform of the music signal. This inspires people looking for applications of wavelet theory in music [63].

2.1.6 Vanishing moments

Not surprisingly, the sparsity property plays an important role in wavelet compression algorithms. As we explain in Section 2.7, it is also the basis for noise reduction by wavelet coefficient thresholding. To create a really sparse representation, we try to make coefficients that live between points of singularities as small as possible. In these intervals of smooth behavior, the signal can be locally well approximated by a polynomial. Therefore, we are interested in polynomials having zero coefficients. If all monomials up to a degree $\tilde{p} - 1$ satisfy

$$\langle x^n, \tilde{\psi} \rangle = 0, \ n = 0, \ldots, \tilde{p} - 1, \tag{2.14}$$

we are sure that the first $\tilde{p}$ terms in a Taylor approximation of an analytic function do not contribute to the wavelet coefficient, provided that there is no singularity in the support of the corresponding dual wavelet. The highest $\tilde{p}$ for which (2.14) holds, is called the (dual) *number of vanishing moments*. For $n < \tilde{p}$, this implies that $x^n \in V_j, \forall j \in \mathbb{Z}$: all polynomials up to a degree equal to the *dual* number of vanishing moments rest within the scaling spaces V_j of the *primal* MRA. Indeed, all detail (wavelet) coefficients of these functions are zero. Each vanishing moment imposes a condition on $\tilde{\psi}$, and so on the wavelet filter $\tilde{g}$. More equations lead to longer filters, and longer filters correspond to basis functions with larger support [158]. This is why we cannot increase the number of vanishing moments *ad infinitum*: the price to pay is loss of locality, the basis functions grow wider and have more chance to get in touch with some of the singularities.

Primal vanishing moments are less important for signal processing applications. We do however prefer at least one vanishing moment, i.e. a zero mean wavelet: this allows for better control of the impact on the output energy of coefficient manipulations. Primal wavelets should be as smooth as possible: each manipulation of a coefficient (for instance thresholding) is actually making a difference between output and input coefficient. After reconstruction, this corresponds to subtracting the corresponding wavelet from the original signal. A non-smooth wavelet shows up in this reconstruction. Vanishing moments are an indication of smoothness, but no guarantee, as illustrates the plot of two biorthogonal wavelets from the Cohen-Daubechies-Feauveau (CDF) [49] family in Figure 2.10. Both primal and dual wavelet have two vanishing moments, but they are clearly not equally smooth. The smoother one is the best candidate for the role of primal (syntheses) wavelet. The other one serves as analysis wavelet.

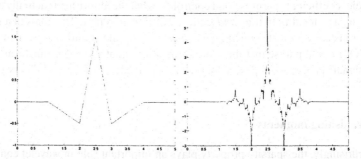

Figure 2.10. Primal and dual wavelet function with two vanishing moments from the Cohen-Daubechies-Feauveau family. The smoother function is preferably used as synthesis (primal) wavelet, and the other one then serves as dual or analysis wavelet.

Projection or function values. One argument for a high number of dual vanishing moments comes from approximation theory. Suppose we have $\tilde{p}$ vanishing moments and $f(x)$ is at least $\tilde{p}$ times continuously differentiable, then the projection

$$P_J f = \sum_{k=1}^{2^J} \langle f, \tilde{\varphi}_{J,k} \rangle \phi_{J,k}$$

of f onto the primal scale space V_J converges to f with an error of order $\mathcal{O}(2^{-J\tilde{p}})$. This means that for proper application a wavelet analysis, one should start the discrete wavelet transform with the scale space projection coefficients $s_{J,k}, k \in \mathbb{Z}$ as input. In real applications, this projection is rarely known, and many algorithms just take the sample values $f_k = f(k/N)$ as input. In a context of noisy data, the wavelet coefficients following from this procedure are often refered to as *empirical wavelet coefficients*. Other authors however use this notion rather to emphasize the observational character of the input. Empirical is then a synonym for noisy. Using sample values instead of projections actually corresponds to replacing the integral of the projection by some function value. This procedure converges, but only at a rate of $\mathcal{O}(2^{-J})$, thereby destroying all benefits from the vanishing moments. Some authors call this approximation "the wavelet crime" [158]. Ingrid Daubechies has constructed orthogonal scaling bases in which the sample value procedure converges to the orthogonal projection as fast as the projection converges to the true function. The corresponding wavelets were called *coiflets*, after Ronald Coifman, who had asked for their construction. Another possibility is to pre-filter the samples, for instance using a quadrature formula to approximate the projection integral [159, 163]. Filtering however changes the statistics of the input, and therefore causes additional problems. Moreover, in many applications, the input is not just the result of an observation at one very short moment or one very narrow spot. On one hand, this fact makes a well founded approximation of the projection even harder. On the other hand, taking these integrated values as imput can be expected to be less harmful than real sample values.

2.1.7 Two-dimensional wavelet transforms

In applications like digital image processing, we need a two-dimensional transform. Although much effort has been and is still being devoted to general, non-separable 2D wavelets [156, 115, 114, 167], this discussion limits itself to separable 2D transforms, i.e. consecutive one-dimensional operations on columns and rows of the pixel matrix. We have two constructions of separable transforms:

1. The square wavelet transform
   The square wavelet transform first performs one step of the transform on all rows, yielding a matrix where the left side contains down-sampled lowpass coefficients of each row, and the right contains the highpass coefficients, as illustrated in Figure 2.11 (a). Next, we apply one step to all columns, this results in four types of coefficients:
   a) coefficients that result from a convolution with $\tilde{g}$ in both directions (HH) represent diagonal features of the image.
   b) coefficients that result from a convolution with $\tilde{g}$ on the columns after a convolution with $\tilde{h}$ on the rows (HL) correspond to horizontal structures.
   c) coefficients from highpass filtering on the rows, followed by lowpass filtering of the columns (LH) reflect vertical information.
   d) the coefficients from lowpass filtering in both directions are further processed in the next step.
   At each level, we have three *components*, *orientations*, or *subbands*: vertical, horizontal and diagonal. If we start with:

$$f(x,y) = \sum_{k \in \mathbb{Z}} \sum_{l \in \mathbb{Z}} s_{J,k,l} \varphi_{J,k}(x) \varphi_{J,l}(y)$$

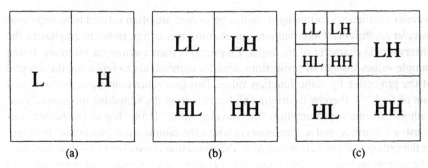

Figure 2.11. A two-dimensional wavelet transform. First we apply one step of the one dimensional transform to all rows (a). Then, we repeat the same for all columns (b). In the next step, we proceed with the coefficients that result from a convolution with $\tilde{h}$ in both directions (c).

the transform decomposes this into:

$$
\begin{aligned}
f(x,y) \;=\; & \sum_{k\in\mathbb{Z}}\sum_{l\in\mathbb{Z}} s_{L,k,l}\varphi_{L,k}(x)\varphi_{L,l}(y) \\
& + \sum_{j=L}^{J-1}\sum_{k\in\mathbb{Z}}\sum_{l\in\mathbb{Z}} w_{j,k,l}^{HH}\psi_{j,k}(x)\psi_{j,l}(y) \\
& + w_{j,k,l}^{LH}\psi_{j,k}(x)\varphi_{j,l}(y) \\
& + w_{j,k,l}^{HL}\varphi_{j,k}(x)\psi_{j,l}(y)
\end{aligned}
$$

2. **The rectangular wavelet transform**

Instead of proceeding with the LL-coefficients of the previous step only, we could also further transform all rows and all columns in each step. This leads to the *rectangular* two-dimensional wavelet transform, illustrated in Figure 2.12. If $w = \tilde{W}y$ is the matrix representation of a 1D wavelet transform, then the rectangular transform, applied to an image I is:

$$\tilde{W}I\tilde{W}^{T}$$

The basis corresponding to this decomposition contains functions that are tensor products of wavelets at *different* scales:

$$\psi_{j,k}(x)\psi_{i,l}(y)$$

Such functions do not appear in the basis of a square wavelet transform.

This alternative not only requires more computation, it is also less useful in applications: in the *square* wavelet transform, the HL and LH components contain more specific information on horizontal or vertical structures.

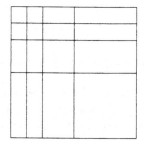

Figure 2.12. Graphical representation of wavelet coefficients after three steps of the rectangular wavelet transform: in each step all rows and all columns are completely further transformed.

2.2 Continuous wavelet transform

So far, this text has been discussing the *Discrete Wavelet Transform* (DWT). In many applications, also in image analysis and even image processing [12, 11], the *continuous* wavelet transform (CWT) plays an important role. Although both are related to a certain degree, the CWT starts from a quite different point of view. There is no multiresolution analysis here, at least not in the sense of the mathematical definition, nor is there any father function involved. The theory immediately introduces a wavelet function $\tilde{\psi}(x)$ and a corresponding wavelet transform $F(a, b)$ of a function $f(x) \in L_2[0, 1]$:

$$F(a, b) = \langle f, \tilde{\psi}_{a,b} \rangle, \tag{2.15}$$

where $a \in \mathbb{R}_0^+, b \in \mathbb{R}$ and

$$\tilde{\psi}_{a,b}(x) = \frac{1}{\sqrt{a}} \tilde{\psi}\left(\frac{x - b}{a}\right).$$

The notion of scale enters into this transform through the *continuous* parameter a.

In principle, we allow all function $\tilde{\psi}(x)$ to play the role of wavelet, provided they guarantee a reconstruction of the input signal through:

$$f(x) = \frac{1}{C_{\tilde{\psi}}} \int_{-\infty}^{\infty} \int_{-\infty}^{\infty} F(a, b) \tilde{\psi}_{a,b}(x) \frac{da\,db}{a^2}. \tag{2.16}$$

So, for invertibility, we need the constant $C_{\tilde{\psi}}$ to be finite. This is the admissibility condition:

$$C_{\tilde{\psi}} = \int_{-\infty}^{\infty} \frac{\tilde{\Psi}(\omega)}{\omega} d\omega < \infty.$$

In this expression, $\tilde{\Psi}(\omega)$ stands for the Fourier transform of $\tilde{\psi}(x)$. In most cases, the admissibility condition is satisfied if the pole in the integrand is neutralized by a zero of $\tilde{\Psi}(\omega)$, this is:

$$\tilde{\Psi}(0) = \int_{-\infty}^{\infty} \tilde{\psi}(x)dx = 0.$$

Functions $\tilde{\psi}(x)$ with zero integral typically show a oscillating behavior, hence the name wavelets.

No other condition rests on the wavelets in a continuous transform. All wavelets from the DWT-theory remain good candidates here, other important functions do not fit into a MRA, but are often used in CWT. Examples are the Morlet wavelet:

$$\tilde{\psi}(x) = e^{i\omega_0 x}e^{-x^2/2\sigma_0^2},$$

and the Mexican hat wavelet:

$$\tilde{\psi}(x) = (1 - x^2)e^{-x^2/2}.$$

This is the second derivative of a Gaussian.

The CWT is highly redundant: it maps a 1D signal into a bivariate function. Obviously, this transform has other applications than the DWT. Whereas the latter appears in fast algorithms for signal- and image *processing* (reconstruction, synthesis), a continuous transform is mainly useful for the *characterization* of signals (analysis). The evolution of $F(a, b)$ as a function of a gives information about smoothness at location b. Loosely spoken, if at location b, the value of $F(a, b)$ increases for small scales, we may expect a short range singularity, such as noise. Large values at coarse scales indicate a long range singularity, typically an important signal feature. A singularity at position b also affects a neighborhood. The evolution of this *cone of influence* across scales a is another regularity indicator. A CWT distinguishes different types of signal singularities, including oscillating singularities [19], and noise. *Uniform* regularity is reflected by the decay of the Fourier transform, but a CWT is able to detect *local* regularity [130, 126]. Section 3.7 discusses how a DWT is the appropriate tool for functions with *global* but *piecewise* regularity properties.

2.3 Non-decimated wavelet transforms and frames

Historically, the continuous wavelet transform came first. The link with filter banks and multiresolution only became clear at the end of the eighties [129]. If the wavelet function $\tilde{\psi}(x)$ fits into a MRA, discretizing the continuous wavelet transform by:

$$a = 2^{-j} \quad \text{and} \quad b = 2^{-j}k$$

leads to the DWT. If there is no MRA however, associated with the wavelet function, this discretization scheme does not allow for a simple reconstruction.

When performing a CWT on a computer, it is common to discretize the location parameter b at sample frequency 2^J (scale 2^{-J}, resolution level J), this is:

$$b_{j,k} = 2^{-J}k,$$

independent from level j. This yields the same number of

$$w_{j,k} = F(a_j, b_{j,k}) = \langle f, \tilde{\psi}_{a_j, b_{j,k}} \rangle$$

at each scale. Therefore, this is an overcomplete data representation. The functions $\tilde{\psi}_{a_j, b_{j,k}}$ do not possibly constitute a basis, because of the oversampling of b. For ease of notation, we write $\tilde{\psi}_{j,k}$ from now on. This set of functions is called a *frame* if

$$\exists A, B \in \mathbb{R}_0^+, \forall f \in L_2[0,1] : A\|f\|^2 \leq \sum_{j,k \in \mathbb{Z}} \langle f, \tilde{\psi}_{j,k} \rangle \leq B\|f\|^2. \qquad (2.17)$$

In that case, we can find functions $\psi_{j,k}$ and reconstruct the input as:

$$f(x) = \sum_{j,k \in \mathbb{Z}} \langle f, \tilde{\psi}_{j,k} \rangle \psi_{j,k} = \sum_{j,k \in \mathbb{Z}} w_{j,k} \psi_{j,k}.$$

The set $\{\psi_{j,k}, j, k \in \mathbb{Z}\}$ shows of course the same degree of redundancy. Figure 2.13 contains a frame in $\mathbb{R}^2$. In this case, the sets $\{\psi_{j,k}\}$ and $\{\tilde{\psi}_{j,k}\}$ coincide. This is called a *tight* frame; in terms of the frame constants, this situation corresponds to the case $A = B$.

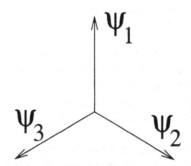

Figure 2.13. An example of a *tight* frame in $\mathbb{R}^2$. It holds for all vectors $f \in \mathbb{R}^2$ that: $f = \sum_{i=1}^3 \langle f, \psi_i \rangle \psi_i$.

A typical example of a wavelet frame follows from a dyadic discretization of the scale parameter a:

$$a_j = 2^{-j}, j = L, \dots, J-1.$$

The frame consists of translations and dyadic dilations of one mother function:

$$\tilde{\psi}_{j,k}(x) = \tilde{\psi}(2^j x - 2^{j-J} k).$$

If the mother wavelet fits into a MRA, the frame coefficients follow from a multiscale filter algorithm, very similar to the fast wavelet transform algorithm, using

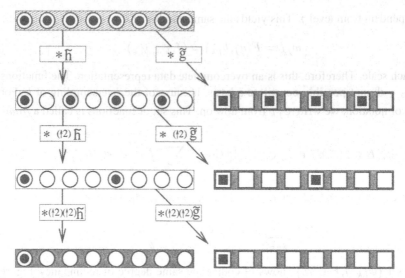

Figure 2.14. The redundant wavelet transform. The points with a black center represent co-efficients that also appear in the decimated transform. To be consistent with this decimated transform, we should make sure that we only combine intermediate results from the original transform in our computation of coefficients "with a black center". To this end, we insert at each level, new zero elements between the filter coefficients of the previous step. This up-sampling operation is represented by (↑2).

filter banks. More precisely, as Figure 2.14 explains, this transform results from omitting the sub-sampling step in a classical wavelet algorithm. Thinking of this transform as an extension of the FWT, we want of course this overcomplete representation to be *consistent* with the decimated version, in the sense that all the decimated coefficients re-appear in our new transform. To compute, for instance, the wavelet coefficients on the one but finest resolution level, we cannot, like in the decimated case, just convolve the scaling coefficients of the previous step with the high frequency filter $\tilde{g}$. If we want to get the original coefficients among our redundant set, we have to skip the extra coefficients of the previous step before the actual convolution. Of course these extra coefficients serve in their turn to complete the redundant set of wavelet coefficients at the given resolution level. A similar procedure is necessary for the computation of the scaling coefficients at the next level. At each level, the number of coefficients to skip, increases as a power of two minus one. As a matter of fact, instead of sub-sampling the coefficients, this alternative introduces up-sampling of the filters $\tilde{h}$ and $\tilde{g}$. Indeed, the wavelet and scaling coefficients at a certain resolution level can be seen as the result of a convolution with filters that are obtained by inserting zeros between the filter coefficients of the previous step. This adaptation preserves the multiresolution character of the wavelet transform: the synthesis frame functions are now:

$$\psi_{j,k}(x) = \psi(2^j x - 2^{j-J} k).$$

Orthogonal wavelet transforms have that $\tilde{\psi}(x) = \psi(x)$. In this redundant scheme, we get a tight frame. This does not mean that all tight frames are built up from an orthogonal transform, as Figure 2.13 illustrates. But the properties of a tight frame are similar to those of an orthogonal basis, just as a general frame recalls the properties of a Riesz basis.

For obvious reasons, the scheme in Figure 2.14 is known as *Non-decimated wavelet transform*, or *Redundant Wavelet Transform*. Other nomenclature includes *Stationary Wavelet Transform*, referring to the translation invariance property. This transform appears in several papers [117, 118, 137, 148, 125], for various reasons, some of which we mention in Section 5.2, while discussing the applicability of this oversampled analysis in noise reduction.

2.4 Wavelet packets

From the algorithmic point of view, there is no reason why the detail coefficients at a given scale in the scheme of Figure 2.7 could not be further processed in the same way as the scaling coefficients. In the Haar transform, taking averages and differences of wavelet coefficients results in basis functions like those shown in Figure 2.15. The coefficients corresponding to these new basis functions may be

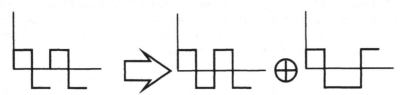

Figure 2.15. Haar wavelet packets: taking averages and differences of wavelet coefficients results in new basis functions. These new coefficients in their turn can be further processed. Further transform of detail coefficients leads to basis functions with wider support. Proceeding with all coefficients at all scales in a Haar transform corresponds to the Hadamard transform.

subject to further transform as well. If we continue this process up to the coarsest scale, we end up with basis functions which all cover the complete interval, thereby losing the locality of a wavelet transform. In the specific case of the Haar filters, this scheme reduces to the so called Hadamard transform.

There is however no need to transform all details or scaling coefficients up to the coarsest scale. In each step and for each subset of coefficients, the algorithm can decide whether or not a filter bank would lead to a more economic data representation [55]. This process is called a *wavelet packets transform* [54]. It intrinsically selects one basis out of a collection of possible bases (often called a library or dictionary). Section 2.9 discusses the use of these principles in the context of noise reduction algorithms.

2.5 Smooth local trigonometric bases

A trigonometric basis as used in the Discrete Fourier Transform, consists of sine and cosine waves: these functions have no local support. To deal with this problem, a classical solution is to break up the input into data blocks and perform a *Short Time Fourier Transform* (STFT) on these data subsets. This approach implicitly creates a periodic extension of all data blocks, thereby introducing singularities at all splitting points (unless data blocks happen to be periodic). Moreover, operating on coefficients in two adjacent blocks probably reveals the location of the boundary between those blocks, since it is unlikely that the operations in both blocks match perfectly. This phenomenon is called the blocking effect.

If the input is even resp. odd around a splitting point κ, an even or odd extension, followed by a periodic extension is much smoother. It is easy to check that

$$f(2\kappa - x) + f(x)$$

is even around κ for every input $f(x)$ and

$$f(x) - f(2\kappa - x)$$

is odd. This procedure can however not be repeated in more than one splitting point, since adding the *mirror function* $f(2\kappa - x)$ has a global effect. Therefore, mirroring should be accompanied with smooth cutoff functions to limit the influence of the mirror function. The total of this pre-processing is called *folding*. Folding input data followed by a STFT corresponds to representing the the data in a basis of smoothly decaying, local trigonometric functions [132, 131, 52].

2.6 Lifting and second generation wavelets

2.6.1 The idea behind lifting

Beside the extension of the *Haar* transform to a filter bank algorithm, there exists another way of generalizing the exploration of intersample correlations: the lifting scheme [164, 160, 161]. Figure 2.16 illustrates the idea. First the data are split into even and odd samples. Both parts are highly correlated. The scheme then *predicts* the odd samples using the information from the even ones. This prediction is called *dual lifting* for reasons explained below. Subtracting this prediction from the actual odd values leads to the detail or wavelet coefficients. Dual lifting for a Haar transform is particularly simple: an odd sample is predicted by the value of its left even neighbor, the difference between them is the wavelet coefficient. As in the filter bank algorithm, these detail coefficients are typically small on intervals of smooth signal behavior. Staying with this one prediction step is however not sufficient to build all types of wavelet transforms. Concentrating on the Haar case, for instance, we see that we did not compute the *average* of the two subsequent samples, only the difference. As a consequence, the *meaning* of the coefficients is different, i.e. the

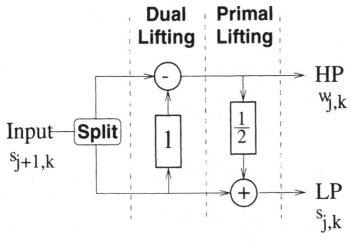

Figure 2.16. Decomposition of a haar transform into lifting steps. The first type of lifting is called *dual lifting* or a *prediction step*. The other type is *primal lifting* or *update*. This diagram is readily invertible by running it in the opposite direction, replacing the split operation by a 'merge' and replacing all plus signs by minus signs and vice versa

basis functions are not those of Figure 2.3. Indeed, the detail coefficient does not indicate how far two input data are below and above the common mean value, but rather how far the second, odd point is from the first, even one. The corresponding detail basis function is a single block, not a block wave, as illustrated in Figure2.17.

Therefore, we want to *update* the *meaning* (interpretation) of the detail coefficient without changing its *value*. We replace the *even* sample by the average of the two consecutive values. Since

$$\text{average} = \frac{\text{even} + \text{odd}}{2} = \text{even} - \frac{\text{even} - \text{odd}}{2} = \text{even} - \frac{\text{difference}}{2},$$

we compute this average by adding an *update* based on the detail coefficients to the even samples. Because this lifting step changes the synthesis or primal wavelet basis function (the 'interpretation' of the coefficient), it is called the primal lifting step. As a consequence, the prediction step is called the dual lifting step.

This idea can be generalized to more complex primal and dual lifting steps. A typical example of a prediction operator is an interpolating polynomial . Figure 2.18 shows a cubic interpolating prediction. The first primal lifting step may be followed by a new dual step and so on. Each step adds more properties — for instance more vanishing moments — to the overall transform: it is a gradual increase in complexity, hence the name lifting. The scheme in Figure 2.19 is the lifted equivalent of a filter bank in Figure 2.5. All classical filter bank algorithms can be decomposed into an alternating sequence of primal and dual lifting steps [62].

This decomposition has several advantages: it generally saves computations, although the order of complexity obviously cannot be better than $\mathcal{O}(N)$ as for the

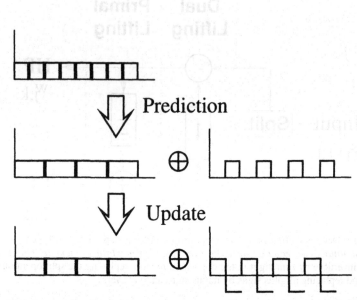

Figure 2.17. Intermediate basis functions in the lifted version of a Haar transform. After the prediction step, we have computed how far the odd samples are from the even ones. This corresponds to extending the even basis functions over the joint even and odd intervals. The wavelets, i.e. the detail basis functions, are just the odd scaling functions at the previous, fine level. These wavelets clearly have no vanishing moments. The update step changes the basis functions corresponding to the odd samples by updating the values of the even samples.

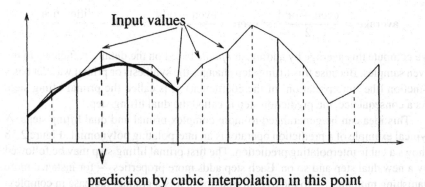

Figure 2.18. A cubic interpolation as a prediction operator. The thin, piecewise linear line links the input data. The bold line is an example of a cubic polynomial, interpolating 4 successive points with even index.

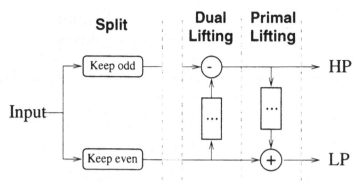

Figure 2.19. Decomposition of a filter bank into lifting steps. The first type of lifting is called *dual lifting* or a *prediction step*. The other type is *primal lifting* or *update*.

classical FWT. Figure 2.19 shows that the result of each dual and primal filtering step with input from one branch is simply added to the other branch. The result of this addition may be stored immediately at the place of the output branch coefficients. The transform is *in place* (computation), it requires no additional working memory. The inverse transform is easy to construct: it uses the same filters in the opposite order and subtracts the result that had been added in the forward transform and vice versa. In the classical filterbank setting, complicated biorthogonality conditions for perfect reconstruction rest on the filter coefficients: these are solved using Fourier techniques. The most important property of lifting is its generality: the dual and primal lifting steps are by no means limited to the classical, linear filter operations. This opens the way to a new, 'second generation' of wavelets. The next sections discuss two examples of these new wavelets.

2.6.2 Subdivision

2.6.3 The integer wavelet transform

In many applications, like digital image processing, the input data are integer numbers. The filters of a wavelet transform are mostly fractions or even irrational numbers, and so is then the output. Performing the transform requires floating point computation *and* storage. The lifting scheme *an sich* does not bring any remedy to this: the coefficients remain the same, regardless of the way of computing, but Figure 2.20 shows that rounding the filter outputs creates a transform that maps integers to integers, called the *Integer Wavelet Transform* [36]. Rounding is not possible in a classical filterbank scheme, since this would destroy perfect reconstruction. In the lifting scheme, the input of each filter operation remains available after this step has been concluded. Going back is always possible by recomputing the filter result.

2.6.4 Non-equidistant data

The lifting philosophy is by no means limited to equidistant samples [162]. The idea of interpolation, for instance, can be extended to an irregular grid, as shows

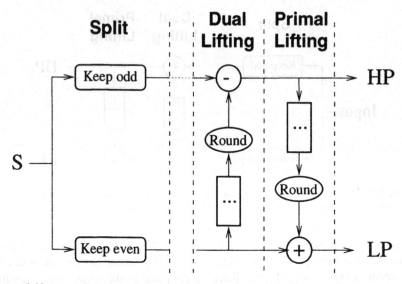

Figure 2.20. Integer wavelet transform.

Fig. 2.21 in the case of linear interpolation. Of course, one could forget about the

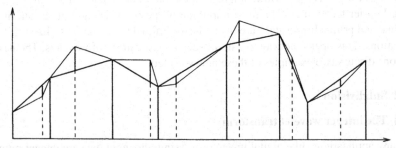

Figure 2.21. Linear prediction operator on an irregular grid.

grid where the input data live on, just treat these points as if they were regular and apply a classical wavelet transform. This does not correspond to reality. For instance, if there is large gap in the measurements, there may be an important difference in signal values in the two end points of this gap, just because they are so far away from each other. If we consider them as samples at uniform distances, it looks as if there is an important singularity at this place. This is obviously a wrong analysis. On the synthesis side we have basis functions with a given smoothness on a regular grid. If we stretch the sample points to fit within the irregular grid, this smoothness is lost. In other words, if we process the wavelet coefficients, the grid irregularity shows up in the result after reconstruction. A scheme that takes into account the grid structure guarantees a smooth reconstruction on this grid [61].

2.7 Noise reduction by thresholding wavelet coefficients

2.7.1 Noise model and definitions

Most noise reduction algorithms start from the following additive model of a discrete signal f of N data points corrupted with noise η:

$$y = f + \eta. \tag{2.18}$$

The vector y represents the input signal. The noise is a vector of random variables, while the untouched values f are a purely deterministic signal. We call N the length of these vectors. Some descriptions start from a full stochastic model, letting the uncorrupted values be an instance from a random distribution. This leads to Bayesian estimators, as we explain later.

We suppose that the noise has zero mean, i.e. $E\eta = 0$ and define

$$Q = E[(\eta - E\eta)(\eta - E\eta)^T] = E\eta\eta^T$$

the covariance matrix of η. On the diagonal we find the variances $\sigma_i^2 = E\eta_i^2$. If this matrix is diagonal, i.e. if $E\eta_i\eta_j = 0$ for $i \neq j$, the noise is called *white* or *uncorrelated*. If all the data points come from the same probability density, we say that the points are *identically distributed*. This implies of course

$$\sigma_i^2 = \sigma^2, \forall i = 1 \ldots N.$$

Noise with constant variance is called *homoscedastic*. Non-homoscedastic noise is heteroscedastic. Homoscadastic, white noise has a simple covariance matrix:

$$Q = \sigma^2 I.$$

Stationarity also involves the correlation between successive observations: the distance between two observations only determines whether and how much these data are mutually dependent. In the special case of second order stationarity, the covariance between two data points only depends on the distance between these two observations. This text mostly assumes second order stationary data. This always includes homoscedasticity.

An important density model is the joint Gaussian:

$$\phi_\eta(\eta) = \frac{1}{(2\pi)^{N/2}\sqrt{\det Q}} \, e^{-\frac{1}{2}\eta^T Q^{-1}\eta}.$$

If Gaussian noise variables are uncorrelated, they are also *independent*. The reverse implication holds for all densities. A classical assumption in regression theory is that of independent, identically distributed noise (i.i.d.). For Gaussian variables this is equivalent with supposing stationary and white noise. Additive, white Gaussian noise is often abbreviated as AWGN.

2.7.2 The wavelet transform of a signal with noise

The linearity of a wavelet transform leaves the additivity of model (2.18) unchanged. We get:

$$w = v + \omega, \tag{2.19}$$

where v is the vector of uncorrupted (untouched, noise-free) wavelet coefficients, ω contains the wavelet transform of the noise and w are the observed wavelet coefficients:

$$
\begin{aligned}
w &= \tilde{W}y, \\
v &= \tilde{W}f, \\
\omega &= \tilde{W}\eta.
\end{aligned}
$$

As before, $\tilde{W}$ is the forward wavelet transform matrix.

With these definitions, it is easy to prove that the covariance matrix of the noise in the wavelet domain $S = \mathrm{E}\omega\omega^T$ equals:

$$S = \tilde{W}Q\tilde{W}^T. \tag{2.20}$$

This equality holds for a general linear transform $\tilde{W}$.

If $\tilde{W}$ is a one-dimensional wavelet transform, this can be interpreted as the *rectangular* wavelet transform of the correlation matrix of the data vector. This should not be confused with the fact that we use a square wavelet transform for the decomposition of 2D data, like images.

If $\tilde{W}$ is orthogonal and $Q = \sigma^2 I$, then we have that $S = \sigma^2 I$. This means that:

Observation 2.1 *An orthogonal wavelet transform of stationary AND white noise is stationary AND white.*

A wavelet transform decorrelates a signal with structures. It leaves uncorrelated noise uncorrelated. Figure 2.22 illustrates what this means for the wavelet transform of a signal with noise. The noise is spread out evenly over all coefficients, and the important signal singularities are still distinguishable from this noisy coefficients. The situation becomes slightly more complicated in the case of non-orthogonal transforms or correlated noise. Chapter 5 discusses these cases. For this and the next two chapters, we work with stationary, white noise and orthogonal transforms.

2.7.3 Wavelet thresholding, motivation

The plot of wavelet coefficients in Figure 2.22 suggests that small coefficients are dominated by noise, while coefficients with a large absolute value carry more signal information than noise. Replacing the smallest, noisy coefficients by zero and a backwards wavelet transform on the result may lead to a reconstruction with the essential signal characteristics and with less noise. More precisely, we motivate this idea by three observations and assumptions:

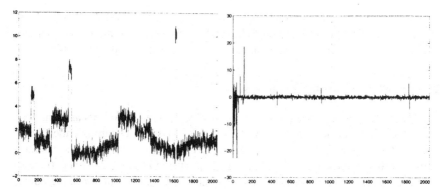

Figure 2.22. Test signal with stationary and white noise (Left) and Haar transform (Right): this is an orthogonal wavelet transform. The noise is spread out equally over all the coefficients, and the important signal singularities are still distinguishable from this noisy coefficients.

1. The decorrelating property of a wavelet transform creates a sparse signal: most untouched coefficients are zero or close to zero.
2. Noise is spread out equally over all coefficients.
3. The noise level is not *too* high, so that we can recognize the signal and the signal wavelet coefficients.

If we replace all coefficients in Figure 2.22 with magnitude below a well chosen *threshold* $\lambda = 1$, we get wavelet coefficients as in Figure 2.23. Figure 2.23(right) shows the corresponding reconstructed signal, which is indeed less noisy than the input. Wavelet thresholding combines simplicity and efficiency and therefore it is

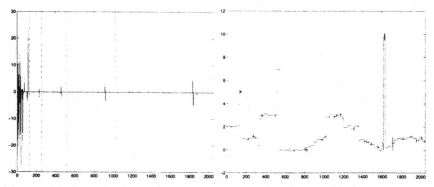

Figure 2.23. Wavelet coefficients after soft-thresholding with threshold $\lambda = 1$ (Left) and reconstruction by inverse Haar transform (Right).

an extensively investigated noise reduction method [70, 76, 71, 43]. The sparsity of representation also motivates the use of wavelets for compression applications and so there are links between compression and noise reduction: wavelet compression techniques show a noise suppressing effect [166] and thresholding based on the technique of context modeling [42, 41] or on the principle of Minimum Description Length (MDL) [97] allows for simultaneous noise reduction and compression.

2.7.4 Hard- and soft-thresholding, shrinking

Until now we have suggested a procedure in which small coefficients are removed, while the others are left untouched. This 'keep-or-kill' procedure is called *hard-thresholding*. Figure 2.24(a) plots the output coefficient versus the input. An alternative for this scheme is soft-thresholding, illustrated in Figure 2.24(b): coefficients above the threshold are shrunk in absolute value. The amount of shrinking equals the threshold value, so that the input-output plot becomes continuous. While at first sight hard-thresholding may seem a more natural approach, the continuity of the soft-thresholding operation has important advantages. In the analysis of algorithms it may be mathematically more tractable. Some algorithms even do not work in combination with hard-thresholding. This is the case for the GCV procedure of Chapter 4. As becomes clear through the next chapters, pure noise coefficients may pass a threshold. In the hard-thresholding scheme, they appear in the output as annoying, spurious 'blips'. Soft-thresholding shrinks these false structures.

A compromise is a more continuous approach as in Figure 2.24(c). It preserves the highest coefficients and has a smooth transition from noisy to important coefficients. Several functions have been proposed [31, 84]. Some of these depend on more than one threshold parameter, others do not introduce any threshold explicitly. An important class of these methods result from a Bayesian modeling of the noise-reduction problem, as we discuss in Section 2.8. In general, these sophisticated shrinking schemes are computationally more intensive. Soft-thresholding is a trade-off between a fast and straightforward method and a continuous approach.

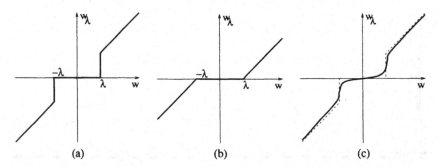

Figure 2.24. Noise reduction by wavelet shrinking. (a) Hard-thresholding: a wavelet coefficient w with an absolute value below the threshold λ is replaced by 0. Coefficients with an absolute value above the threshold are kept. (b) Soft-thresholding: Coefficients with magnitude above the threshold are shrunk. (c) a more sophisticated shrinking function.

2.7.5 Threshold assessment

A central question in many threshold procedures is how to choose the threshold. As we explain in Chapter 3, a threshold is a trade-off between closeness of fit and smoothness. A small threshold yields a result close to the input, but this result may still be noisy. A large threshold on the other hand, produces a signal with a lot of zero wavelet coefficients. This sparsity is a sort of smoothness: the output has a simple, smooth representation in the chosen basis. Paying too much attention to smoothness however destroys some of the signal singularities, in image processing, it may cause blur and artifacts.

Literature contains a bunch of papers devoted to this problem of threshold selection. The next two chapters describe a method that looks for the minimum mean square error threshold: this threshold minimizes the error of the result as compared with the noise-free data. Since these data are unknown, the error cannot be computed or minimized exactly. Estimating the minimum is a topic in some papers [78, 106, 136, 135].

The *universal threshold* [71, 81] pays attention to smoothness rather than to minimizing the mean square error. We discuss this well known threshold in Chapter 3. This threshold comes close to the minimax threshold, i.e. the threshold that minimizes the worst case mean square error in a typical function space [80].

Other methods consider wavelet coefficient selection as an example of (multiple) hypothesis testing [2, 146, 145].

2.7.6 Thresholding as non-linear smoothing

There is a formal way to demonstrate that thresholding is a particular example of a more general class of smoothing algorithms. These algorithms typically look for a compromise between closeness of fit and smoothness. Smoothness, or sparsity, can be expressed by some measure of 'entropy', which should be minimized. On the other hand, for closeness of fit, the algorithms use an error 'energy' term, mostly the norm of the difference between input and output. A smoothing parameter λ takes care of the compromise between these two and the algorithm minimizes:

$$\|w_\lambda - w\|^2 + \lambda \mathcal{E}(w_\lambda),\qquad(2.21)$$

where $\mathcal{E}(w_\lambda)$ is the entropy of the output.

For this entropy there exist many expressions.

1. Using the ℓ_2 norm

$$\mathcal{E}(w_\lambda) = \|w_\lambda\|^2$$

leads to a linear shrinking operation:

$$w_\lambda = \frac{1}{1+\lambda}w.$$

By linearity of the wavelet transform, this would correspond to shrinking the untransformed data, unless we leave the scaling coefficients untouched, but even then, this operation does not make so much sense: the ℓ_2 norm is measure of energy, not sparsity or entropy.

We could make the contributions level-dependent:

$$\mathcal{E}(\boldsymbol{w_\lambda}) = \sum_{j=L}^{J-1} \gamma_j \|\boldsymbol{w_{\lambda,j}}\|^2,$$

Taking $\lambda_j = \lambda\gamma_j$, we see this is equivalent to a level-dependent shrinking:

$$\boldsymbol{w_{\lambda,j}} = \frac{1}{1+\lambda_j}\boldsymbol{w_j}.$$

The smoothing parameters could be chosen to optimize the mean square error of the result, although in practical applications, this error cannot be computed exactly. We could limit the possibilities *a priori* to:

$$\lambda_j = 0 \quad \text{for} \quad j = L, \ldots, K-1,$$
$$\lambda_j = \infty \quad \text{for} \quad j = K, \ldots, J-1$$

This means: keep the first coefficients, *in casu* the coefficients at low resolution levels, and throw away the next ones at finer scales. Keeping the *first* is always a linear operation. Now K is acting as smoothing parameter.

2. Keeping the *largest* coefficients is a *non-linear* operation, it corresponds to minimizing:

$$\sum_{i=1}^{N} \left(|w_{\lambda i} - w_i|^k + \lambda^k x(w_{\lambda i}) \right).$$

In this expression the *label* $x(w)$ is zero if the output $w_{\lambda i}$ is exactly zero, and one in all other cases. For all k, this leads to hard-thresholding:

$$w_{\lambda i} = w_i \quad \text{if} \quad |w_i| \geq \lambda$$
$$w_{\lambda i} = 0 \quad \text{if} \quad |w_i| < \lambda$$

For $k = 2$, this reduces to the form of (2.21), where the entropy is the number of non-zero coefficients:

$$\mathcal{E}(\boldsymbol{w_\lambda}) = N_1 = \#\{i = 1, \ldots, N | w_{\lambda i} \neq 0\},$$

and of course we use λ^2 instead of λ in (2.21), but this is only a matter of notation.

3. Soft-thresholding follows if we use the ℓ_1 norm as measure of sparsity. We minimize

$$\sum_{i=1}^{N}\left((w_{\lambda i}-w_i)^2+2\lambda|w_{\lambda i}|\right),$$

this means that the entropy in (2.21) equals:

$$\mathcal{E}(\boldsymbol{w}_\lambda)=2\sum_{i=1}^{N}|w_{\lambda i}|.$$

2.8 Other coefficient selection principles

Coefficient thresholding is only one example of a wider class of algorithms, which proceed in three steps:

1. Wavelet transform of the input.
2. *Manipulation* of the *empirical* wavelet coefficients.
3. Inverse wavelet transform of the modified coefficients.

Most important and characteristic is of course the second step. The manipulation aims at a noise reduction, without losing too much signal information. In most algorithms, this manipulation depends on a *classification* of a coefficient, which is often binary: a coefficient is either *noisy* or *relatively noise-free* and important. To distinguish between these classes, we need a *criterion*, a sort of threshold on a measure of *regularity*. There are essentially two types of models on which this measure of regularity is based: Bayesian and non-Bayesian models.

The former type thinks of the uncorrupted coefficients $\boldsymbol{V}$ as an instance from a density function $f_{\boldsymbol{V}}(v)$, so we get a fully random model:

$$\boldsymbol{W}=\boldsymbol{V}+\boldsymbol{N},$$

where $\boldsymbol{N}$ is the noise. In principle, these methods compute the posterior density for the noise-free coefficients from Bayes' rule:

$$f_{\mathbf{V}|\mathbf{W}}(v|w)=\frac{f_{\mathbf{V}}(v)f_{\mathbf{W}|\mathbf{V}}(w|v)}{f_{\mathbf{W}}(w)},\tag{2.22}$$

which allows, for instance, to estimate the underlying 'true' coefficients by the posterior mean:

$$\hat{v}=\mathrm{E}(\boldsymbol{V}|\boldsymbol{W}).$$

Depending on the chosen density functions $f_{\mathbf{V}}(v)$ and $f_{\mathbf{W}|\mathbf{V}}(w|v)$, this mostly leads to shrinking rules [46, 48, 57, 151, 171, 4, 155, 104]. Chapter 6 discusses an example of such a method.

The other type of classification considers the noise-free signal as a deterministic member of a smooth function space. These methods try to understand how signal regularity can be observed from wavelet coefficients:

1. As seen before, coefficients with a large magnitude are important. This is however not the only possible measure of regularity.
2. Simple thresholding is a very local approach: each coefficient is classified independent from its neighbors. It does not take into account the correlations between different coefficients, especially across scales. Other methods [180] are based on the assumption that regular signal or image features show correlated coefficients at different scales, whereas irregularities due to noise do not. These algorithms compute the *correlation* between coefficients at successive scales.
3. A third class of methods is based on the characterization of the Lipschitz or Hölder regularity of a function by its (continuous) wavelet transform [126, 128]. These methods look at the *evolution* of the coefficients across the different scales to distinguish regular from noisy contributions. As mentioned before, a regular image or signal singularity has a long-term range and therefore the corresponding wavelet coefficients at coarse scales are large. Noise on the contrary, is local and therefore its singularities have larger coefficients at finer scales.

At first sight, these methods may seem more heuristic, but they avoid the possibly extensive use of hyperparameters in a Bayesian approach. Those parameters have to be chosen, often on a heuristic basis.

The algorithms based on Lipschitz characterization actually use a continuous wavelet transform, or a sampled version of it, i.e. some sort of non-decimated transform. This is a overcomplete data representation. The essential signal information is captured by the local coefficient extrema on each scale. The evolution through scales allows to distinguish signal extrema from noise extrema. These local extrema suffice to reconstruct the input signal, apart from some pathological cases [126, 128]. The reconstruction scheme, proposed by Mallat and collaborators is a time-consuming iterative process.

2.9 Basis selection methods

Wavelet coefficient manipulation methods proceed within one wavelet basis, or one pair of dual bases. More adaptive methods build up the basis in which the given signal or image is processed. The objective is to make the signal fit as well as possible into this self-constructed basis. The method uses an overcomplete set of functions φ_i, in the sense that a given signal can be expressed as a linear combination of more than one subset of these 'library' or 'dictionary' of functions. Well structured libraries lead to fast algorithms (typically order $\mathcal{O}(N \log N)$) for best basis selection, i.e. finding the basis in which the coefficients a in the decomposition

$$f = \sum_i a_i \phi_i$$

has minimal entropy. As in Section 2.7.6, the concept of entropy has several possible definitions. The ℓ_1-norm is one of these, and another well known example is:

$$\mathcal{E}(a) = \sum_i \frac{-|a_i|}{\|a\|_2} \log \frac{|a_i|}{\|a\|_2}.$$

One could also consider the sum of squares of coefficients above a given threshold as an entropy measure [75].

As explained in Section 2.4, the wavelet packet transform belongs to this methods of best basis selection [54]. Other methods use different types of functions, like local trigonometric functions [50].

If the input data y are noisy, the noise can be eliminated by a decomposition

$$y = \sum_{i=1}^{m} a_i \phi_i + R^{(m)},$$

so that:

$$\frac{1}{2}\|R^{(m)}\|_2^2 + \lambda \mathcal{E}(a) \tag{2.23}$$

is as small as possible. The result is a trade-off between a close approximation of the input and a sparse representation. In this objective function, λ plays the role of smoothing parameter. The idea is that noise cannot be sparsely represented in any basis from the library. Some variants include:

1. *Basis pursuit* [45] uses the ℓ_1-norm as entropy. The objective function (2.23) then reduces to the form of a linear programming problem. Moreover, this expression has the same form as the one leading to soft-thresholding in the fixed basis setting. This motivates a choice of the smoothing parameter λ similar to the universal threshold.
2. *Matching pursuit* [127] is a greedy algorithm. In each step, it looks for the function from the library with the highest correlation with the residual after the previous step.
3. *Best Orthogonal Basis* [55] is limited to orthogonal bases. Extensions are in [51, 50].

2.10 Wavelets in other domains of application

Noise reduction by wavelet thresholding is an example of *non-parametric regression*. Similar techniques are used for *density estimation*, but the settings for this problem are different, as we briefly discuss in Section 5.8. Apparently, the wavelet world gets more and more penetrated with statisticians. Other applications in statistics [1] include time series analysis (stochastic processes) [138, 139, 140], change point analysis [146, 145, 14, 150, 175], and inverse problems [111, 5, 72].

Other operations from signal processing [165] and more general system theory as well as identification problems belong to popular subjects of investigation for wavelet methods. The input may come from various domains, like geology, geography, or financial data.

Among all possible fields of application for wavelet based methods, digital image processing is probably the most visible or visual one. Most problems and solutions from one-dimensional signal processing have an equivalent in image processing, but the real challenge in this field is of course developing algorithms which are not mere more-dimensional versions of classical digital signal processing operations.

Not only wavelet based noise reduction schemes are applicable to images. The analysis of wavelet extrema by Mallat and collaborators, discussed in the previous section, also opens the way to multiresolution image contrast enhancement methods [121, 116, 142] and the decorrelating properties of a wavelet transform (the sparsity properties) are the basis for many image and video compression applications. Compression is closely related to image processing and can be seen as an example of approximation. Approximation theory is another field of application and, as becomes clear from subsequent chapters, its results may be interesting to explain wavelet based smoothing algorithms.

The flexibility of the lifting scheme for the construction on all types of lattices turns out to be useful in computer graphics and geometrical modeling [60, 164].

Another important domain of application is numerical analysis [159]. For the solution of partial differential equations, for instance, wavelet methods may serve as preconditioners [58, 88, 89].

A domain which is at first sight a bit further away from the material in this text is theoretical and mathematical physics.

2.11 Summary and concluding remarks

Wavelet theory combines the following properties:

1. A wavelet transform has a *decorrelating* property. A wavelet decomposition leads to a *sparse* representation. This is useful in compression applications and is a basis for noise reduction algorithms by wavelet thresholding.
2. Wavelet theory naturally supports the idea of *multiresolution*. Since a lot of phenomena in nature have a multiscale character, the ability to analyze and process data level-dependent is interesting in many applications. Images are a typical example of multiscale data: they contain information, objects of different scales.
3. Wavelet basis functions are *local* in time/space and frequency (scale). Manipulating the corresponding coefficients has a local effect: this allows good control on the effect of these manipulations.
4. A wavelet decomposition is a linear transform with linear complexity. This allows fast algorithms.
5. Orthogonality or bi-orthogonality in a Riesz-basis guarantee numerically well conditioned transforms.
6. The variety of wavelet basis functions and corresponding filters allows for each application an ideal choice of working basis.

7. Wavelet methods are based on a sometimes difficult, but nice mathematical background.

Wavelet methods for non-parametric regression or noise reduction, can be catalogued according to several characteristics:

1. The transform technique
   a) in a basis (fast wavelet transform, ridgelets, local cosine bases, ...)
   b) in a frame (e.g. non-decimated wavelet transform)
   c) best basis selection (e.g. using wavelet packets)
2. the wavelet properties being exploited
   a) sparsity (threshold algorithms)
   b) multiresolution (interscale correlations; tree-structured algorithms)
   c) intrascale correlations (e.g. using a Markov Random Field model, like in Chapter 6)
   d) locality
3. the model being used
   a) Bayesian (full Bayes or empirical Bayes)
   b) non-Bayesian (with elements from approximation theory, function classes, smoothing)
   The objectives:
   a) do we need a *smooth* output or
   b) is it more important that the output error is as small as possible?

Evidently, the application at hand determines which method is prefered. Other issues, like speed of computations, are also considered.

The next two chapters concentrate on the *sparsity* and *locality* to motivate wavelet thresholding for noise reduction. In Chapter 5, the *multiresolution* character of the transform turns out to be useful when dealing with less standard, academic situations of noisy signals. If the data live on an irregular grid, the lifting scheme comes in. This happens in Chapter 7. This lifting scheme has the following properties:

1. The lifting scheme for performing a wavelet transform speeds up computations, although the general order of complexity remains of course linear. Moreover, all computations are in place.
2. The basic ideas are easier to understand and implement: it is a more intuitive approach and does not require any Fourier techniques.
3. The inverse transform is trivial to construct from the data flow picture.
4. The lifting approach is more generic. It allows for extensions to non-equispaced samples and integer transforms. In the former case, lifting guarantees a smooth reconstruction. Stability however, does not seem to be guaranteed, as Chapter 7 points out.

3

The minimum mean squared error threshold

This chapter investigates the mean squared error as a criterion for selecting an optimal soft threshold. In applications like image processing, it is often objected that this expression of the error does not always correspond to a more subjective experience of quality. Our visual system, for instance, is much more sensitive to contrast than is expressed by a mean squared error. Nevertheless, even in the image processing world, definitions of signal-to-noise ratio, based on mean squared errors, are commonly used.

On the other hand, the material of this chapter is not limited to a specific application. Moreover, the ideas can easily be extended to representations, different from wavelet bases. The only thing we need, is a data set where a few coefficients carry a large proportion of the information, so that an algorithm can "throw away" an important part of the data without losing substantial information. The decorrelating property of a wavelet transform provides us with this sparsity . This chapter is therefore based on this decorrelation.

In the first section, we introduce the mean square error as a function of the threshold value, and examine its typical shape. Next, we focus on the threshold that minimizes this this objective function. We try to understand how it behaves asymptotically, i.e. if the number of data N tends to infinity. We operate in two steps: first we study piecewise polynomials, and second we turn to the general piecewise smooth function case. The outcome of this study reveals an interesting similarity to the well-known *universal threshold*. Section 3.4 resumes the principal properties of this often used threshold. The mean square error has already been analyzed from different points of view [40, 93, 95, 94, 30, 133]. Apart from Section 3.4, the analysis in this chapter is original material.

3.1 Mean square error and Risk function

3.1.1 Definitions

In the previous chapter, we already learnt that a threshold can be seen as a smoothing parameter: it controls the compromise between goodness of fit and smoothness of approximation. In this context, smoothness should be interpreted as sparsity: we try to find a sparse data set, close to the noisy input.

The ultimate objective is of course an approximation of the noise-free data. While balancing between closeness of fit and sparsity, the *best* compromise minimizes the error of the result as compared with these unknown, uncorrupted data.

If y_λ is the output of the threshold algorithm with some threshold value λ and f is the vector of untouched data, the remaining noise on this result equals $\eta_\lambda = y_\lambda - f$, and the mean squared error (MSE) is then defined as:

$$R(\lambda) = \text{MSE}(\lambda) = \frac{1}{N}\|\eta_\lambda\|^2. \tag{3.1}$$

As the notation indicates, the MSE, $R(\lambda)$, is a function of the threshold value λ. It is also a random variable, because it depends on the noise. The expected value of this error is called the *risk*-function.

The main challenge with this MSE as an objective function is the fact that in real applications, it can never be computed exactly: its definition uses the value of the exact, unknown data f. In practical situations, this MSE has to be estimated. As the next chapter discusses, GCV is such an estimator.

A common definition of signal-to-noise ratio (SNR) is based on this notion of MSE:

$$\text{SNR}(\lambda) = 10 \cdot \log_{10} \frac{\|f\|^2}{\|\eta_\lambda\|^2} = 10 \cdot \log_{10} \frac{\|f\|^2/N}{R(\lambda)}. \tag{3.2}$$

An alternative is the peak signal-to-noise ratio, which is equal to the previous one, up to constant, depending on the uncorrupted data:

$$\text{PSNR}(\lambda) = 10 \cdot \log_{10} \frac{(\max f)^2/N}{R(\lambda)} = \text{SNR}(\lambda) + 10 \cdot \log_{10} \frac{(\max f)^2}{\|f\|^2}. \tag{3.3}$$

Both SNR and PSNR are expressed in deciBels (dB) .

An orthogonal wavelet transform $\tilde{W}$ preserves the ℓ_2-norm, and so:

$$R(\lambda) = \frac{1}{N}\|\omega_\lambda\|^2,$$

where $\omega_\lambda = w_\lambda - v = \tilde{W}(y_\lambda - f)$. From now on, we do all our computations in the wavelet domain. If the transform is biorthogonal, there is no exact equivalence with the data domain. Nevertheless, computation and minimization in terms of wavelet coefficients seems to give satisfactory results, and several reasons could explain this: Riesz-bounds guarantee a nearly equivalent norm. Moreover, since MSE does not correspond exactly to a human perception of quality, the question arises whether MSE in the original data domain is always a better measure than MSE in the wavelet domain. In image processing applications, for instance, we view the image in the pixel domain, but we do not look at an image as a matrix of pixels. Since our visual system seems to work on a multiscale basis, a norm based on a multiresolution decomposition might be a better expression of visual quality. Further illustrations show that there is no need for expressing norms in the original data domain. This preserves us from applying an inverse wavelet transform every time we want to evaluate the quality of a result. An inverse wavelet transform is only necessary to compute the eventual output of the algorithm.

3.1.2 Variance and bias

The input wavelet coefficients are unbiased estimates of the noise-free coefficients:

$$E w = E \tilde{W} y = \tilde{W} E y = \tilde{W} f = v$$

but the variance of this "estimation" is too high. Replacing the smallest coefficients with zero reduces the variance, at the cost of an increasing bias:

$$\mathrm{bias}^2(\lambda) \quad := \quad \frac{1}{N} \| E w_\lambda - v \|^2 \tag{3.4}$$

$$\mathrm{variance}(\lambda) \quad := \quad \frac{1}{N} E \| w_\lambda - E w_\lambda \|^2. \tag{3.5}$$

Then it holds that:

$$ER(\lambda) \quad = \quad \frac{1}{N} \| E w_\lambda - v \|^2 + \frac{1}{N} E \| w_\lambda - E w_\lambda \|^2 \tag{3.6}$$

$$\mathrm{Risk} \quad = \qquad \mathrm{bias}^2 \qquad + \qquad \mathrm{variance}$$

The most reliable method to remove *all* noise is just removing everything:

$$\lim_{\lambda \to \infty} \mathrm{variance}(\lambda) = 0.$$

If all coefficients are removed, there is no variance anymore, all the noise has gone, but so has the signal: the bias equals the total energy of the noise-free input:

$$\lim_{\lambda \to \infty} \mathrm{bias}^2(\lambda) = \frac{1}{N} \| v \|^2$$

Figure 3.1 shows a typical behavior of these functions. The minimum risk threshold is the best compromise (in ℓ_2) between variance and bias.

In Chapter 2, we motivated thresholding as a smoothing algorithm: our objective was to find a compromise between closeness of fit and smoothness. Next, we introduced the notion of MSE and Risk to define the *best* compromise. This risk function

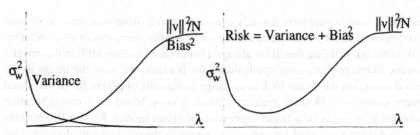

Figure 3.1. Typical behavior of bias and variance as a function of the threshold value. Thresholding introduces bias, but reduces variance. The best compromise minimizes the risk.

is a sum of two effects: variance and bias, and the minimum risk threshold is again the best compromise between these two. We already mentioned that the MSE or the risk function cannot be possibly computed in real applications. Unlike smoothness and closeness of fit, both variance and bias are unknown in practice, but a solution with small variances is probably smooth whereas a close fit generally shows little bias. This link is implicitly present when estimating the optimal threshold with cross validation, as becomes clear in the next chapter.

3.2 The risk contribution of each coefficient (Gaussian noise)

This section puts some elementary calculations together. The results are necessary for the next sections. From now on, we assume that the input noise is Gaussian and we call:

$$\phi(\omega) = \frac{1}{\sqrt{2\pi}\sigma}e^{-\omega^2/2\sigma^2},$$

$$\Phi(\omega) = \int_{-\infty}^{\omega} \phi(u)du.$$

Every classical, linear wavelet transform preserves the normality of a density. If the input noise is not Gaussian, the density of the wavelet coefficients, if at all computable in practice, would depend on the type of wavelets being used.

Some of the following results also appear in different papers like [76]. A first lemma gives an expression for the bias of one coefficient $w = v + \omega$ (the notation omits the index of the coefficient).

Lemma 3.1

$$
\begin{aligned}
E\omega_\lambda &= \sigma^2[\phi(\lambda - v) - \phi(\lambda + v)] + \lambda[\Phi(\lambda - v) - \Phi(\lambda + v)] \\
&\quad + v[1 - \Phi(\lambda - v) - \Phi(\lambda + v)].
\end{aligned}
\tag{3.7}
$$

The proof is by simple calculations, using the fact that a Gaussian distribution satisfies the following differential equation:

$$\omega\phi(\omega) = -\sigma^2\phi'(\omega). \tag{3.8}$$

We denote by

$$r(v,\lambda) = \mathrm{E}(w_\lambda - v)^2 \tag{3.9}$$

the contribution of coefficient w to the total risk function. Using Equation (3.8) and partial integration leads to

$$\int \omega^2\phi(\omega)d\omega = -\omega\sigma^2\phi(\omega) + \sigma^2 \int \phi(\omega)d\omega,$$

which allows to conclude, after some calculation, that:

Lemma 3.2

$$
\begin{aligned}
r(v,\lambda) \;=\; & \left[2(\sigma^2 + \lambda^2) - v^2\right] + \left[\Phi(\lambda - v) + \Phi(\lambda + v)\right]\left(v^2 - \sigma^2 - \lambda^2\right) \\
& -\sigma^2\left[(\lambda - v)\phi(\lambda + v) + (\lambda + v)\phi(\lambda - v)\right].
\end{aligned}
\tag{3.10}
$$

Plots of this contribution as a function of the threshold λ for various values of v show that coefficients with little information ($v \approx 0$) are best served with large thresholds, whereas important coefficients (v large) prefer little thresholding. The overall optimal threshold is the best compromise between these two.

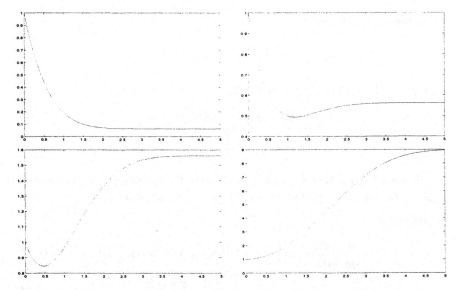

Figure 3.2. Contribution of individual coefficients to the total risk as a function of the threshold value. Small values of v (upper left) prefer large thresholds, because bias is small. Large values of v (lower right) would cause considerable bias if the threshold gets large.

To find the minima, we compute the derivative. Again a trivial computation leads to:

Lemma 3.3

$$\frac{\partial r}{\partial \lambda}(v, \lambda) = 2\lambda \left[1 + \Phi(-v - \lambda) - \Phi(-v + \lambda)\right] - 2\sigma^2 \left[\phi(v + \lambda) + \phi(-v + \lambda)\right].$$

(3.11)

The proof uses the fact that

$$\frac{\partial}{\partial \lambda} E(w_\lambda - v)^2 = E \frac{\partial}{\partial \lambda}(w_\lambda - v)^2.$$

An important case is that of a coefficient without any information. It turns out that if $v = 0$, the derivative $\frac{\partial r}{\partial \lambda}$ is always negative (see Figure 3.2, upper left). If $\lambda \to \infty$, the derivative approaches zero, but it remains negative. This means that the optimal threshold for this zero coefficient equals infinity. This is confirmed by the following asymptotic behavior:

Lemma 3.4

$$\frac{\partial r}{\partial \lambda}(0, \lambda) \sim -4\sigma^4 \frac{\phi(\lambda)}{\lambda^2}.$$

(3.12)

Proof:
From the previous lemma, we see that:

$$\frac{\partial r}{\partial \lambda}(0, \lambda) = 4\lambda[1 - \Phi(\lambda)] - 4\sigma^2 \phi(\lambda).$$

Three times De L'Hôpital's rule shows that:

$$\lim_{\lambda \to \infty} \frac{\lambda[1 - \Phi(\lambda)] - \sigma^2 \phi(\lambda)}{\frac{\phi(\lambda)}{\lambda^2}} = -\sigma^4.$$

□

To get an idea of how $\frac{\partial r}{\partial \lambda}$ behaves more generally, we compute the derivative of this expression with respect to the uncorrupted coefficient value v:

Lemma 3.5

$$\frac{\partial}{\partial v}\left(\frac{\partial r}{\partial \lambda}(v, \lambda)\right) = 2v\left[\phi(v + \lambda) + \phi(-v + \lambda)\right]$$

(3.13)

$$\begin{cases} \geq 0 & \text{if } v \geq 0, \\ \leq 0 & \text{if } v \leq 0. \end{cases}$$

(3.14)

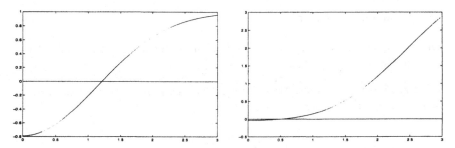

Figure 3.3. Derivative of the risk in a given coefficient with respect to the threshold value as a function of the noise-free coefficient value. Left: $\lambda = 0.5\,\sigma$. For coefficients approximately below $1.1\,\sigma$, this threshold is too low. Right: $\lambda = 2\sigma$. This threshold is too large for coefficients above $0.5\,\sigma$. The distribution of noise-free coefficients determines the optimal threshold.

Consequently, for a given threshold, $\frac{\partial r}{\partial \lambda}$ has a minimum for $v = 0$. Figure 3.3 shows $\frac{\partial r}{\partial \lambda}$ as a function of v for two different values of λ. The plot on the left-hand side corresponds to a threshold $\lambda = 0.5\,\sigma$. This threshold is too small for all coefficients smaller than approximately $1.1\,\sigma$. If we choose $\lambda = 2\sigma$, only coefficients below $0.5\,\sigma$ find this value too small. The value of the optimal threshold depends on how the noise-free coefficients are distributed: the sparser the representation is, the larger the optimal threshold will be. Indeed, if the proportion of small coefficients increases, the threshold should be large, because all these small coefficients prefer large thresholds. The next sections try to find an asymptotic behavior for this optimal threshold. We assume that generating more samples from a given signal on a continuous line, introduces more redundancy in the information. This causes more sparsity in the wavelet representation. We expect that the optimal threshold increases as the number of samples N grows.

3.3 The asymptotic behavior of the minimum risk threshold for piecewise polynomials

3.3.1 Motivation

The next chapter shows that, in minimum risk sense, the GCV-method asymptotically yields the optimal threshold. This property motivates the use of GCV in a threshold assessment procedure. For the proof of this asymptotic optimality, we need to know how the optimal threshold itself behaves if the number of samples $N \to \infty$. This section assumes that the samples come from a piecewise polynomial on $[0, 1]$:

$$y_i = f(i/N) + \eta_i,$$

where $f(t)$ is a piecewise polynomial and $t \in [0, 1]$. No real signal is of course a perfect piecewise polynomial, but typical signals are piecewise smooth. Section 3.6 investigates how the threshold behaves in this more general case.

3.3.2 Asymptotic equivalence

Before studying the asymptotics of the minimum risk threshold, we recall the definition of asymptotic equivalence:

Definition 3.1 *Two functions $f(x)$ and $g(x)$ are said to be asymptotic equivalent for $x \to \infty$, i.e. $f(x) \sim g(x)$, if and only if*

$$\lim_{x \to \infty} \frac{f(x)}{g(x)} = 1.$$

The study of the asymptotics of the minimum risk threshold uses a couple of properties of this notion:

Lemma 3.6 *Let $x \to \infty$. If $f(x) \sim g(x)$, we have:*

1. $f(x)h(x) \sim g(x)h(x)$.
2. *If $f(x) \sim h(x)$, then $h(x) \sim g(x)$.*
3. *If $h(x) = o(f(x))$, then $f(x) \pm h(x) \sim g(x)$.*
4. *If*

$$\lim_{x \to \infty} f(x) \neq 1 \neq \lim_{x \to \infty} g(x),$$

and both functions are differentiable, then

$$\log f(x) \sim \log g(x).$$

Proof:

1. Trivialiter
2. $\displaystyle \lim_{x \to \infty} \frac{h(x)}{g(x)} = \lim_{x \to \infty} \frac{h(x)}{f(x)} \frac{f(x)}{g(x)}$,

   and we may split the limit of this product, since both factors have a limit.
3. This is actually a special case of the previous statement.

$$\lim_{x \to \infty} \frac{f(x) \pm h(x)}{g(x)} = \lim_{x \to \infty} \frac{f(x)}{g(x)}\left(1 + \frac{h(x)}{f(x)}\right) = \lim_{x \to \infty} \frac{f(x)}{g(x)} \cdot \lim_{x \to \infty} \left(1 + \frac{h(x)}{f(x)}\right)$$

4. For $\displaystyle \lim_{x \to \infty} f(x) = \infty$, we use De L'Hôpital's rule:

$$\lim_{x \to \infty} \frac{\log f(x)}{\log g(x)} = \lim_{x \to \infty} \frac{f'(x)/f(x)}{g'(x)/g(x)} = \lim_{x \to \infty} \frac{f'(x)}{g'(x)} \lim_{x \to \infty} \frac{g(x)}{f(x)}$$

$$= \lim_{x \to \infty} \frac{f(x)}{g(x)} \lim_{x \to \infty} \frac{g(x)}{f(x)} = 1.$$

For a finite limit, we do not need the differentiability:

$$\lim_{x\to\infty} \frac{\log f(x)}{\log g(x)} = \frac{\lim_{x\to\infty} \log f(x)}{\lim_{x\to\infty} \log g(x)} = \frac{\log\left(\lim_{x\to\infty} f(x)\right)}{\log\left(\lim_{x\to\infty} g(x)\right)} = 1$$

$\square$

We remark that the inverse implication of the last statement definitely does not hold: if $\log f(x) \sim \log g(x)$, $f(x)$ and $g(x)$ may be not asymptotically equivalent. For instance, if $\log f(x) = x + \sqrt{x}$ and $\log g(x) = x$, then

$$\frac{f(x)}{g(x)} = e^{\sqrt{x}} \not\to 1.$$

3.3.3 The asymptotic behavior

For the piecewise polynomial case, we assume that the wavelet analysis has more vanishing moments than the highest degree of the polynomials. As a consequence, wavelet coefficients are zero if they do not correspond to a basis function which interferes with a singularity . We assume that the number of singularities is finite on $[0, 1]$.

We then have the following theorem for the asymptotic behavior of the minimum of $ER(\lambda) = E\|y_\lambda - f\|^2$:

Theorem 3.1 *Suppose $f(x)$ is a piecewise polynomial on $[0; 1]$, $v_i := v_{j,k}, i = 2, \ldots N = 2^J, j = 0, \ldots, J - 1, k = 1, \ldots, 2^j$ are the wavelet coefficients of the orthogonal projection of f on V_J. Call $w_i = v_i + \omega_i$ the noisy wavelet coefficients and $R(\lambda)$ the soft-threshold MSE-function as defined in (3.1). If λ^* minimizes $ER(\lambda)$, then for $N \to \infty$:*

$$\lambda^* \sim \sqrt{2 \log N} \sigma \qquad (3.15)$$

Proof:
We suppose that the wavelet transform is orthogonal, so the problem model in the wavelet domain is the same as in the input (time or space) domain:

$$w = v + \omega,$$

where ω is i.i.d. noise with variance σ^2. A wavelet coefficient v_i or w_i corresponds to a basis function $\psi_{j,k}$ at resolution level j and place k. Sometimes we use the double index notation for these coefficients: $v_i = v_{j,k}$ and $w_i = w_{j,k}$.

We call:

$$
\begin{aligned}
I_0 &= \{i = 1, \ldots, N | v_i = 0\} \\
I_1 &= \{i = 1, \ldots, N | v_i \neq 0\} \\
M_0 &= \#I_0 \\
M_1 &= \#I_1
\end{aligned}
$$

M_0 and M_1 of course depend on N. Since $f(t)$ is a piecewise polynomial, at each level only a constant number of coefficients is not exactly zero. The total number of non-zero coefficients is proportional to the number of levels:

$$M_1 \sim \log N,$$

and so:

$$\frac{M_1}{N} \to 0.$$

Using the notation from the previous section, we may write:

$$ER(\lambda) = \frac{1}{N} \sum_{i=1}^{N} r(v_i, \lambda).$$

And so:

$$ER'(\lambda) = \frac{M_0}{N} \frac{\partial r}{\partial \lambda}(0, \lambda) + \frac{1}{N} \sum_{i \in I_1} \frac{\partial r}{\partial \lambda}(v_i, \lambda).$$

Lemma 3.3 shows that $\frac{\partial r}{\partial \lambda}(0, \lambda) < 0$. Call $I_1' \subset I_1$ the indices of the non-zeros for which $\frac{\partial r}{\partial \lambda}(v_i, \lambda^*)$ is negative. These indices belong to the smaller coefficients. The indices of the large coefficients are in $I_1'' = I_1 \setminus I_1'$. We define

$$\begin{aligned} M_1' &= \#I_1' \\ M_1'' &= \#I_1''. \end{aligned}$$

We now write the equation for λ^*: $ER'(\lambda^*) = 0$ or:

$$-M_0 \frac{\partial r}{\partial \lambda}(0, \lambda^*) - \sum_{i \in I_1'} \frac{\partial r}{\partial \lambda}(v_i, \lambda^*) = \sum_{i \in I_1''} \frac{\partial r}{\partial \lambda}(v_i, \lambda^*). \tag{3.16}$$

We consider this equation as an equality of two functions of N, and let $N \to \infty$. Both sides of this equation have all positive terms. We now investigate these both sides.

We know that

$$v_i = v_{j,k} \approx \sqrt{N} \int_{\mathbb{R}} f(t)\psi_{j,k}(t)dt,$$

where the integral is a constant value which does not depend on N, so $v_i = \mathcal{O}(\sqrt{N})$, and we are looking for a λ^* which does not increases faster. This means that M_1'' is a non-decreasing function of N: if $N \to \infty$, ever more coefficients are classified as large, since all non-zero coefficients grow at least as fast as the optimal threshold. On the other hand, $M_1'' \leq M_1$ does not increase too fast. If v_i grows faster than λ^* for increasing N, the right-hand side of (3.16) behaves like $M_1'' 2\lambda^*$, as follows from letting $v \to \infty$ in Lemma 3.3.

Lemma 3.5 says that

$$\forall i \in I_1' : \left| \frac{\partial r}{\partial \lambda}(v_i, \lambda^*) \right| \leq \left| \frac{\partial r}{\partial \lambda}(0, \lambda^*) \right|.$$

Moreover $M_1' \leq M_1 \sim \log N$, so $\frac{M_1'}{M_0} \sim \frac{\log N}{N} \to 0$. From this, we may conclude that the sum $\sum_{i \in I_1'}$ in Equation (3.16) can be neglected.

We have:

$$\frac{M_0}{M_1''} \sim \frac{2\lambda^*}{-\frac{\partial r}{\partial \lambda}(0, \lambda^*)}.$$

If $N \to \infty$, the left-hand side grows like $N/\log N \to \infty$. The right-hand side is an increasing function of λ^*: it is easy to verify (from the proof of Lemma 3.4) that

$$-\frac{\partial^2 r}{\partial \lambda^2}(0, \lambda) = -4\left[1 - \Phi(\lambda)\right] \leq 0,$$

so the denominator is a positive, decreasing function, while the numerator is positive and increasing. To make this right-hand side grow to infinity, we need $\lambda^* \to \infty$. Therefore, we can use Lemma 3.4 and get the following asymptotic equation:

$$\frac{M_0}{M_1''} \sim \frac{2\lambda^* \lambda^{*2}}{4\sigma^4 \phi(\lambda^*)},$$

$$\frac{\sigma^3 M_0}{M_1''} \sim \sqrt{\frac{\pi}{2}} \, e^{\lambda^{*2}/2\sigma^2} \lambda^{*3},$$

$$3\log\sigma + \log M_0 - \log M_1'' \sim \frac{\lambda^{*2}}{2\sigma^2} + 3\log\lambda^* + \frac{1}{2}\log\frac{\pi}{2}.$$

The left-hand side depends on N, the right-hand side depends on λ^*. We keep the essential on both sides:

$$\log M_0 \sim \frac{\lambda^{*2}}{2\sigma^2},$$

$$\log(N - M_1) \sim \frac{\lambda^{*2}}{2\sigma^2},$$

$$2\sigma^2 \log N \sim \lambda^{*2}.$$

□

3.3.4 An example

Figure 3.4 shows the plot of a piecewise, linear polynomial. This function is sampled, and transformed into wavelet domain, using the orthogonal Daubechies wavelets with two vanishing moments. We then compute numerically the minimum of $ER(\lambda)$ for different sample rates, and $\sigma = 1$. These values are listed in Table 3.1 and plotted in Figure 3.5.

Both table and figure illustrate that indeed

$$\lambda^* \approx K + \sqrt{2 \log N}\sigma = K + \sqrt{2 \log 2}\sqrt{J},$$

where K is constant and $2^J = N$.

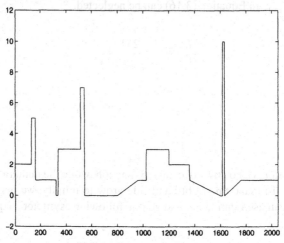

Figure 3.4. An example of a piecewise polynomial: in this case, all pieces are linear or constant.

3.3.5 Why does the threshold depend on the number of data points?

To engineers it might look strange that the optimal threshold depends on the number of data points. They object that the threshold should not change by putting two signals together?

First, this objection does not correspond to the philosophy behind this asymptotic analysis: we do not join two signals, but merely take more samples from one function on a given interval. Second, as Table 3.1 illustrates, we note that $\sqrt{2 \log N}$ is only a very weak dependence.

And third, there is a comprehensive explanation for this behavior. Adding more samples enhances redundancy in the signal: there is less new information in new samples than there was in the first samples. In wavelet domain, this means that the number of important coefficients is hardly growing, and all information remains concentrated in a limited number of coefficients. If we suppose that the transform is normalized, the magnitude of these large coefficients should increase, since more samples mean a higher total energy (2-norm of the data vector) and this energy is preserved by the wavelet transform, while all nearly zero coefficients hardly take any of it. On the other hand, the noise variance in all coefficients remains σ^2 all the time. If the threshold would be independent of N, say $\lambda = k\sigma$, then the relative

J	$N = 2^J$	λ^*
7	128	1.00564887172677
8	256	1.12338708120546
9	512	1.25150560042977
10	1024	1.40212517387946
11	2048	1.55837014248141
12	4096	1.75818138547032
13	8192	1.94789562337191
14	16384	2.13051380127175
15	32768	2.30620858768226
16	65536	2.47521770450833
17	131072	2.63789870923179
18	262144	2.79627013633284
19	524288	2.95567796055755
20	1048576	3.11403413842011

Table 3.1. Minimum risk threshold for the piecewise polynomial in Figure 3.4 as a function of the number N of samples.

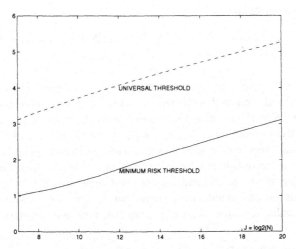

Figure 3.5. Plot of minimum risk threshold for the signal in Figure 3.4 as a function of the binary logarithm J of the number of samples (full line). De dashed line is a plot of the predicted equivalence: $\sqrt{2 \log 2} \sqrt{J}$. The plot seems to confirm this asymptotic behavior.

number of purely noise coefficients which passes the threshold would converge to $P(|Z| > k)$, Z being a standard normal variable. So, the total number of noise coefficients would be proportional to N. Since the number of important coefficients is approximately a constant, the reconstruction would become noisier. Therefore, it is better to let the threshold increase slowly to catch all noise coefficients, while leaving the faster growing signal coefficients intact.

3.4 Universal Threshold

Of course, the formula of the asymptotic behavior of the minimum risk threshold does not tell everything about the actual, optimal threshold value. This actual value depends on all coefficients, while the asymptotic formula only depends on N and σ. We did not say that one should use this asymptotic formula as a real threshold. Nevertheless, the value

$$\lambda_{\text{UNIV}} = \sqrt{2 \log N} \sigma \qquad (3.17)$$

is well known in wavelet literature and it is used as a threshold value, not only as an asymptotic equivalence. This is the so-called *universal threshold*. This name reflects the idea that this threshold is "valid" for all signals with length N, provided that these signals are "sufficiently" smooth: it is a general threshold value. Donoho, Johnstone and collaborators have proven a lot of optimality properties for this choice. We briefly discuss some of these ideas.

3.4.1 Oracle mimicking

Wavelet coefficient shrinking, especially hard-thresholding, can be seen as an example of selective wavelet reconstruction, i.e. based on a given regularity *criterion* some of the coefficients are preserved and others are removed to reconstruct the signal. For thresholding, the measure of regularity is simply the coefficient magnitude. If the underlying uncorrupted coefficient were known, one could of course find the best possible selection. If the objective is to minimize the risk, it is easy to prove that one should select the coefficients for which $|v_i| > \sigma$ and omit the others. All this is of course a completely irrealistic situation: if we knew the untouched v_i, we would not even consider reconstruction by *noisy* coefficient selection. One could imagine an *oracle* telling us whether the uncorrupted coefficients are above σ but not the exact values. This remains an unrealistic dream, but can serve as a benchmark: it leads to the best possible selection. With respect to this, there are two important results [76]:

1. Within a logarithmic factor, (optimal) wavelet coefficient selection performs essentially as well as any piecewise polynomial and spline method: if SW stands for selective wavelet reconstruction, and PP for piecewise polynomial reconstruction, we have that

$$\text{Risk}(\text{SW}) \leq (A \log N + B)(\text{Risk}(\text{PP}))$$

2. Again within a logarithmic factor wavelet thresholding using the universal threshold performs as well as optimal selective wavelet reconstruction:

$$\text{Risk}(\lambda_{\text{UNIV}}) \leq (2 \log N + 1)(\text{Risk}(\text{SW}) + \frac{\sigma^2}{N})$$

3.4.2 Minimax properties

In a certain sense, the previous result also holds in the opposite direction: there is no threshold that in all cases essentially performs better than the universal threshold. More precisely, let λ_{mM} be the minimax threshold, i.e. the largest threshold that minimizes the maximum relative risk with respect to the optimal selection risk. Then it holds that this risk ratio also behaves like $2 \log N$ and the minimax threshold itself is asymptotically

$$\lambda_{\text{mM}} \sim \sqrt{2 \log N} \sigma.$$

3.4.3 Adaptivity, optimality within function classes

The previous two results relate the performance of universal soft-thresholding to the ideal oracle coefficient selection. This is a relative result: basically it states that *if* optimal coefficient selection performs well, thresholding performs nearly as well. This leaves us with the question when selective wavelet reconstruction is a good method for noise reduction. Clearly, we may expect good performance for signals with a sparse wavelet representation. These are typically piecewise smooth signals. To characterize this piecewise smoothness we cannot use the normal concept of C^α (Lipschitz or Hölder; see Definition 3.2) functions, since one singularity destroys the overall smoothness. On the other hand, the L_p spaces may be too general and contain really non-smooth functions. Section 3.7.2 introduces and briefly discusses the concept of Besov spaces. For the moment, we just mention that a Besov space $B_{p,q}^s$ contains piecewise smooth signals and the parameters p, q, s measure different aspects of this smoothness.

If a function f lies in such a space $\mathcal{F}$, the universal threshold risk is guaranteed to come within a logarithmic factor of the minimax risk, i.e. the risk of the estimator which minimizes the worst case risk within the function space:

$$\sup_{f \in \mathcal{F}} \text{Risk}(\lambda_{\text{UNIV}}) \leq C \log N \cdot \inf_{\hat{f}} \sup_{f \in \mathcal{F}} \text{Risk}(\hat{f}), \tag{3.18}$$

where the infimum is taken over all possible estimators $\hat{f}$. The universal threshold comes that close without knowing the exact smoothness parameters p, q, s, whereas the optimal estimator clearly depends on these values. This is why the threshold procedure is called adaptive to unknown smoothness [71].

3.4.4 Smoothness

A Besov space is of course associated with a corresponding norm. This Besov norms measures the smoothness of a function, thereby being "flexible" with singularities: one isolated singularity poses no problem for Besov-smoothness. It turns out that the reconstruction using a universal threshold with high probability (tending to one if $N \to \infty$) is at least as smooth as the untouched signal [71].

This smoothness guarantee is not available for thresholds designed for minimum risk optimality, like SURE and GCV (see next chapter). The output of these algorithms indeed often shows noisy "blips". In Chapter 5 we propose some modifications to the GCV-threshold algorithm to get rid of these annoying false structures.

3.4.5 Probabilistic Upper bound

Since $\|0\|_{\mathcal{F}} = 0$ in every smoothness space $\mathcal{F}$, the previous result implies that if $f = 0$, the reconstruction is the zero function with high probability. There is another way of explaining this constation. From classical extreme value theory we have the following theorem [71, 119]:

Theorem 3.2 *Let* $\{X_k\}_{k \in \mathbb{N}}$ *be an i.i.d. sequence with common distribution function* $F_X(x)$ *and let* $M_N = \max\{X_k | k = 1, \dots, N\}$, *then for a real sequence* λ_N:

$$\lim_{N \to \infty} P(M_N \leq \lambda_N) = e^{-a} \quad \Leftrightarrow \quad \lim_{N \to \infty} N(1 - F_X(\lambda_N)) = a. \qquad (3.19)$$

For $\lambda_N = \sqrt{2 \log N} \sigma$ and $F_X(x) = \Phi(x)$, we can use

$$1 - \Phi(x) \sim \sigma^2 \frac{\phi(x)}{x} \qquad x \to \infty, \qquad (3.20)$$

to find that:

$$\lim_{N \to \infty} N\left(1 - \Phi(\sqrt{2 \log N} \sigma)\right) = \lim_{N \to \infty} N\sigma^2 \frac{1}{\sqrt{2\pi}\sigma} \frac{1}{N\sqrt{2 \log N}} = 0.$$

and so:

$$\lim_{N \to \infty} P(M_N \leq \sqrt{2 \log N} \sigma) = 1.$$

This means that with probability increasing to one, the universal threshold λ_N removes all coefficients that are purely noise. On the other hand, suppose we have a slower growing threshold μ_N and call $\nu_N = \mu_N / \lambda_N$, so that

$$\lim_{N \to \infty} \nu_N < 1,$$

then

$$\lim_{N\to\infty} N\left(1 - F_X(\mu_N)\right) = \lim_{N\to\infty} \frac{e^{\log N} \sigma^2 e^{-(\nu_N \lambda_N)^2/2\sigma^2}}{\sqrt{2\pi}\sigma\nu_N\lambda_N}$$

$$= \lim_{N\to\infty} \frac{e^{\log N} e^{-\nu_N^2 \log N}}{2\sqrt{\pi \log N}\,\nu_N}$$

$$= \lim_{N\to\infty} \frac{N^{1-\nu_N^2}}{2\nu_N\sqrt{\pi \log N}} > 0$$

All slower growing thresholds lack this probabilistic upper bound property. Because of this property, the universal threshold procedure is sometimes called a noise *removal* or *de-noising* technique. Unlike noise *reduction* by minimizing the MSE of the output, the universal threshold does not look for a compromise between noise and bias, but rather removes (asymptotically almost surely) all the noise while preserving as much as possible from the underlying data. Instead of balancing between the two objectives, smoothness comes first, closeness of fit comes next.

3.4.6 Universal threshold in practice

Figure 3.6 shows this universal threshold at work. We add artificial white noise ($\sigma = 0.2$) to test signal with $N = 2048$ data points. Only the finest five levels are thresholded. Using the universal threshold ($\lambda_{\text{UNIV}} = 0.7810$) yields indeed a smoother result than the minimum MSE threshold ($\lambda_{\text{MSE}} = 0.2656$).

In an image processing context, smoothness means blur. Figure 3.7 shows that the universal threshold, without further modifications to the algorithm, is certainly not appropriate for image de-noising. The minimum MSE threshold performs better, but still does not satisfy. Several further modifications to ameliorate this result are discussed in the subsequent chapters.

3.5 False Discovery Rate (FDR)

Section 3.3.5 explained intuitively why the mean risk threshold should depend on the number of wavelet coefficients. Both the minimax threshold and the universal threshold depend on the data size as well.

This observation is related to the notion of *False Discovery Rate* (FDR) in multiple hypothesis testing [23]: testing whether N values are 'significantly different from zero' with a fixed significance level α leads to an average of αM_0 false rejections of the $H_{0,i}$ hypothesis: 'value i is essentially zero'. Here M_0 is the number of 'essentially zero' values. In many applications, such as wavelet thresholding, $M_0 = \mathcal{O}(N)$. Simultaneous inference of different values should limit the total number of type I errors, i.e. the total number of erroneously rejected $H_{0,i}$.

A straightforward approach, known as the Bonferroni test procedure simply considers the N hypotheses independently at a significance level of α/N. This procedure ensures that the overall probability of having even one type I error is less than

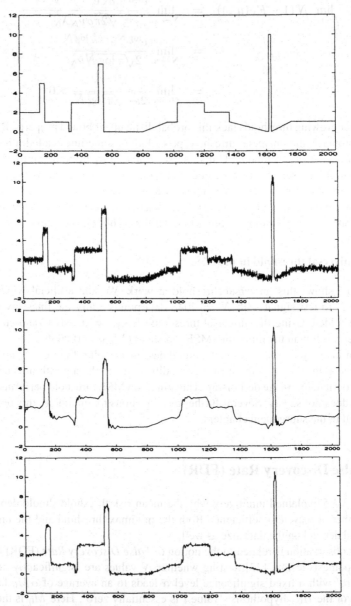

Figure 3.6. Universal threshold at work. The first plot shows a test signal. Next is the same signal with additive, Gaussian i.i.d. noise. The first reconstruction is by thresholding the finest five resolution levels, using a universal threshold ($\lambda_{\mathrm{UNIV}} = 0.7810$). The second reconstruction uses the minimum MSE threshold ($\lambda_{\mathrm{MSE}} = 0.2656$). The universal threshold is larger, and so creates a smoother reconstruction, but it also introduces more bias. We use orthogonal Daubechies wavelets with three vanishing moments here.

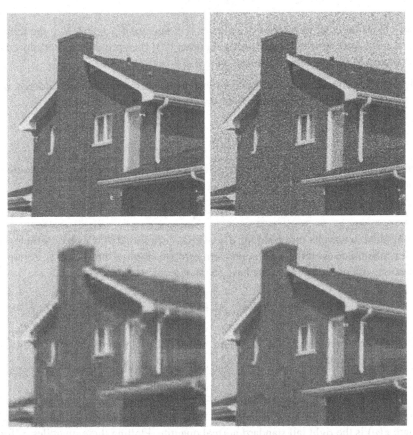

Figure 3.7. Thresholding for images. The image on lower left was obtained by the universal threshold. Smoothness means blur here. The minimum MSE threshold gives a better result (lower right). The wavelet basis is Daubechies' orthogonal one with three vanishing moments, and we process the three finest resolution levels only. The image has 256 by 256 pixels.

α. This procedure is conservative: for large N, the significance criterion becomes extremely strict, leaving numerous non-zero values undetected.

As a matter of fact, in applications like wavelet coefficient selection, it is more important to limit the *relative* number of type I errors. More concrete, we define the False Discovery Rate as

$$FDR = E\frac{N_{1,0}}{N_1},$$

where N_1 is the total number of discoveries (i.e. rejected $H_{0,i}$) and $N_{1,0}$ the number of false discoveries. The following theorem [23] then provides a procedure to guarantee a FDR lower than a chosen, global q^*.

Theorem 3.3 *Consider N independent hypothesis testings H_i with ordered p-values $P_{(i)}, i = 1, \ldots, N$ and let N_1 be the largest index i satisfying*

$$P_{(i)} \leq \frac{i}{N}q^*.$$

Rejecting $H_{0,(i)}, i = 1, \ldots, N_1$ (i.e. rejecting those hypotheses with p value smaller than $P_{(N_1)}$) leads to:

$$FDR \leq q^*.$$

Applied to wavelet thresholding, discoveries correspond to wavelet coefficients with a magnitude essentially different from zero, i.e. above a threshold [2]. Assuming noise with normal density, we have (applying double sided tests):

$$P_{(i)} = 2\left[1 - \Phi_\sigma(|w|_{(i)})\right] = 2\left[1 - \Phi_1\left(\frac{|w|_{(i)}}{\sigma}\right)\right].$$

Using the monotonicity of the (standard nomal) distribution $\Phi_1(x)$ the condition $P_{(i)} \leq \frac{i}{N}q^*$ reduces to:

$$|w|_{(i)} \geq \sigma z\left(\frac{iq^*}{2N}\right),$$

where $z(p)$ is the right tail standard normal quantile. Plotting these quantiles z_i for $p = \frac{iq^*}{2N}$ and looking for the largest index N_1 for which $|w|_{(i)} \geq \sigma z_i$ leads to the (hard) threshold:

$$\lambda_{FDR,q^*} = \sigma z\left(\frac{N_1 q*}{2N}\right).$$

Note that the plots of $|w|_{(i)}$ and σz_i may have more than one crossing. The threshold follows from the last one.

This threshold combines a data driven approach with the objective of smoothness. Smoothness in this case means sparsity: the procedure explicitly aims at minimizing the number of noisy coefficients passing the threshold. The notion of sparsity can be formalized using so called ℓ_p-balls, as Section 3.7.3 explains. Not surprisingly, FDR thresholds show similar adaptivity results as the universal threshold. Emphasizing smoothness of course tends to more bias than a minimum risk approach: risk is a trade-off between smoothness and closeness of fit. Again, the application determines what is preferable. In statistical inference, it is often important to avoid type I errors, since these errors may lead to wrong conclusions. In regression however, type II errors (erroneously accepting the zero hypothesis) may be equally important. Type II errors introduce bias. In an image processing context, bias is blur.

3.6 Beyond the piecewise polynomial case

We now return to the minimum risk threshold. Theorem 3.1 investigated the asymptotic behavior of the minimum risk threshold λ^* for piecewise polynomials. We would like to generalize this result to general piecewise smooth functions. The proof of the polynomial theorem introduced the idea that the optimal threshold is the best compromise between coefficients with a large uncorrupted value, for which this threshold is already beyond the optimum, ($\frac{\partial r}{\partial \lambda}(v_i, \lambda^*) > 0$) and small coefficients, for which the optimal threshold could be larger ($\frac{\partial r}{\partial \lambda}(v_i, \lambda^*) < 0$). This distinction implicitly divides the coefficients into two groups, but we did not compute the boundary v^* between them, because for piecewise polynomials, we can count on the important group of coefficients exactly equal to zero.

For general piecewise smooth functions, none of the coefficients is exactly zero, and therefore we want to have an idea for which and for how many coefficients a given threshold λ^* is too large or too small. This could give us an impression of the behavior of the optimal compromise.

3.6.1 For which coefficients is a given threshold too large/small?

From Lemma 3.3, we learn that:

$$\frac{\partial r}{\partial \lambda}(v, \lambda) = 0 \Leftrightarrow \frac{\sigma^2}{\lambda} = \frac{1 + \Phi(-v - \lambda) - \Phi(-v + \lambda)}{\phi(v + \lambda) + \phi(-v + \lambda)}. \tag{3.21}$$

We now consider this as an equation in v and look for a lower bound for its solution v^* as a function of λ. Lemma 3.5 says that for a fixed threshold value, $\frac{\partial r}{\partial \lambda}(v, \lambda)$ is a monotonically increasing function of v if $v > 0$, and this guarantees that Equation (3.21) has at most one solution.

Let $G(v, \lambda)$ be the right-hand side of Equation (3.21):

$$G(v, \lambda) = \frac{1 + \Phi(-v - \lambda) - \Phi(-v + \lambda)}{\phi(v + \lambda) + \phi(-v + \lambda)}.$$

It is trivial to see that

$$\lim_{v \to \infty} G(v, \lambda) = \infty.$$

If we find a value v^0 for which $G(v^0, \lambda) \le \frac{\sigma^2}{\lambda}$, we may conclude that the solution v^* of (3.21) satisfies $v^* \ge v^0$.

We evaluate

$$G\left(\frac{\sigma^2}{\lambda}, \lambda\right) = \sigma \, \frac{2 - \Phi_1(u + 1/u) - \Phi_1(u - 1/u)}{\phi_1(u + 1/u) + \phi_1(u - 1/u)},$$

in which $u = \lambda/\sigma$ and ϕ_1, Φ_1 are standard normal density and distribution:

$$\phi_1(x) = \frac{1}{\sqrt{2\pi}} \, e^{-x^2/2}.$$

The next, technical section argues that:

$$u \, \frac{2 - \Phi_1(u + 1/u) - \Phi_1(u - 1/u)}{\phi_1(u + 1/u) + \phi_1(u - 1/u)} \le 1 \tag{3.22}$$

for all $u \ge 1.7815$, and so

$$G\left(\frac{\sigma^2}{\lambda}, \lambda\right) \le \frac{\sigma^2}{\lambda}$$

if $\lambda \ge 1.7815 \, \sigma$.

This allows us to formulate the following theorem:

Theorem 3.4 *If $\lambda \ge 1.7815 \, \sigma$, and v^* satisfies*

$$\frac{\partial r}{\partial \lambda}(v^*, \lambda) = 0$$

then

$$v^* \ge \frac{\sigma^2}{\lambda}. \tag{3.23}$$

Figure 3.8 shows a numerical computation of the curve $v^*(\lambda)$. It demonstrates that we have found a sharp lower bound.

In the upcoming analysis we need the fact that $\frac{\partial r}{\partial \lambda}(v, \lambda)$ is convex as a function of v for $|v| \le \sigma^2/\lambda$. Therefore, we formulate an additional lemma:

Lemma 3.7 *If $\lambda \ge \sigma$, we have for $v \le \sigma^2/\lambda$:*

$$\frac{\partial^2}{\partial v^2}\left(\frac{\partial r}{\partial \lambda}\right) \ge 0. \tag{3.24}$$

Proof:
From Lemma 3.5 we compute

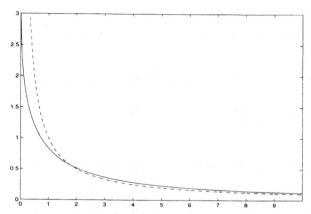

Figure 3.8. Full line: plot of v^* as a function of λ, where v^* satisfies $\frac{\partial r}{\partial \lambda}(v^*, \lambda) = 0$. Dashed line: plot of σ^2/λ. In this example we put $\sigma = 1$.

$$\frac{\partial^2}{\partial v^2}\left(\frac{\partial r}{\partial \lambda}\right) = 2\phi(v+\lambda)\left[1 - \frac{v(v+\lambda)}{\sigma^2}\right] + 2\phi(v-\lambda)\left[1 - \frac{v(v-\lambda)}{\sigma^2}\right]. \quad (3.25)$$

The factor

$$1 - \frac{v(v-\lambda)}{\sigma^2}$$

is positive on

$$\left[\frac{\lambda - \sqrt{\lambda^2 + 4\sigma^2}}{2}, \frac{\lambda + \sqrt{\lambda^2 + 4\sigma^2}}{2}\right],$$

which contains the interval $[0, \lambda]$.

Since $v \leq \sigma^2/\lambda$ and $\lambda \geq \sigma$ by assumption, we have $v \leq \lambda$, and so:

$$\frac{\partial^2}{\partial v^2}\left(\frac{\partial r}{\partial \lambda}\right) \geq 2\phi(v+\lambda)\left[1 - \frac{v(v+\lambda)}{\sigma^2} + 1 - \frac{v(v-\lambda)}{\sigma^2}\right]$$

$$= 4\phi(v+\lambda)\left[1 - \frac{v^2}{\sigma^2}\right].$$

From $v \leq \sigma^2/\lambda$ and $\lambda \geq \sigma$, it also follows that $v \leq \sigma$, and so we know that this expression is positive. $\qquad\square$

Corollary 3.1 *For $\lambda \to \infty$,*

$$\frac{\partial r}{\partial \lambda}\left(\frac{\sigma^2}{2\lambda}, \lambda\right)$$

is negative and tends to 0, but not faster than

$$-2\sigma^4 \frac{\phi(\lambda)}{\lambda^2}$$

Proof:

This follows from the asymptotic behavior of $\frac{\partial r}{\partial \lambda}(0, \lambda)$ in Lemma 3.4, the fact that

$$\frac{\partial r}{\partial \lambda}(\frac{\sigma^2}{\lambda}, \lambda) \leq 0,$$

which follows from Theorem 3.4, and from the previous lemma, stating that $\frac{\partial r}{\partial \lambda}$ is convex between 0 and σ^2/λ. □

3.6.2 Intermediate results for the risk in one coefficient

This leaves us with the question to prove the inequality in (3.22). This section is purely technical, and may be skipped for understanding the rest of this chapter.

Call

$$H(u) = u \, \frac{2 - \Phi_1(u + 1/u) - \Phi_1(u - 1/u)}{\phi_1(u + 1/u) + \phi_1(u - 1/u)}. \tag{3.26}$$

The plot of this function in Figure 3.9 seems to confirm that indeed $H(u) < 1$ for

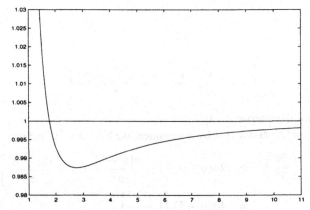

Figure 3.9. Plot of function $H(u)$, defined in (3.26). Important to note is that this function is smaller than 1 for $u > 1.7815$.

$u > 1.7815$. To make sure that this remains true for higher values of u, we start with the following lemma:

Lemma 3.8 *The function*

$$H_0(x) = \frac{1 - \Phi_1(x)}{\phi_1(x)} \tag{3.27}$$

tends to zero for $x \to \infty$ and decreases monotonically for all positive x.

Proof:

The computation of $\lim_{x \to \infty} H_0(x)$ is straightforward, using De L'Hôpital's rule.

Next, it can be verified that $H_0(x)$ satisfies the first order differential equation:

$$H_0'(x) = xH_0(x) - 1,$$

and so

$$H_0''(x) = xH_0'(x) + H_0(x).$$

Suppose $H_0'(x) > 0$. Since $H_0(x) > 0$, this means that $H_0''(x) > 0$. So, $H_0'(x)$ would be positive and increasing. This conflicts with the limit of $H(x)$ to be zero. □

As a consequence of this lemma, and since it follows from $a/b \leq c/d$ that $a/b \leq (a+c)/(b+d) \leq c/d$, we have that for positive u:

$$u\frac{1 - \Phi_1(u + \frac{1}{u})}{\phi_1(u + \frac{1}{u})} \leq u\frac{\left(1 - \Phi_1(u + \frac{1}{u})\right) + \left(1 - \Phi_1(u - \frac{1}{u})\right)}{\phi_1(u + \frac{1}{u}) + \phi_1(u - \frac{1}{u})} \leq u\frac{1 - \Phi_1(u - \frac{1}{u})}{\phi_1(u - \frac{1}{u})}.$$

It is easy to verify that both the left and the right-of this inequalities tend to one if $u \to \infty$, and so, we may conclude that

$$\lim_{u \to \infty} H(u) = 1. \tag{3.28}$$

We now use the fact that

$$\phi_1\left(u - \frac{1}{u}\right) = e^2 \phi_1\left(u + \frac{1}{u}\right)$$

to rewrite $H(u)$ as:

$$H(u) = \frac{u}{e^2 + 1} \frac{2 - \Phi_1(u + \frac{1}{u}) - \Phi_1(u - \frac{1}{u})}{\phi_1(u + \frac{1}{u})}.$$

From this expression we calculate:

$$H'(u) = -u - \frac{e^2 - 1}{e^2 + 1}\frac{1}{u} + \left(u + \frac{1}{u} - \frac{1}{u^3}\right)H(u).$$

This allows us to prove that:

Lemma 3.9 *The function $H(u)$, as defined in Equation (3.26), satisfies:*

$$H(u) \leq 1, \quad \forall u \geq \sqrt{\frac{e^2 + 1}{2}} \approx 2.048. \tag{3.29}$$

Proof:

Suppose $H(u) \geq 1$, then

$$H'(u) \geq \frac{2}{e^2 + 1}\frac{1}{u} - \frac{1}{u^3}.$$

This expression is positive for $u \geq \sqrt{\frac{e^2+1}{2}}$, which means that $H(u)$ would increase and its limit could never become one. □

3.6.3 Piecewise smooth functions

If the noise-free signal is an exact polynomial, and if the multiresolution analysis has sufficiently many vanishing moments, the signal can be written as a linear combination of scaling functions at an arbitrary resolution. This means that all detail coefficients, i.e., the wavelet coefficients, are exactly zero. We used this property to describe what happens with *piecewise* polynomials. To investigate piecewise smooth functions, we follow the same way: we start with properties for wavelet coefficients of functions with a certain degree of smoothness. Of course, these coefficients will not be exactly zero, but smooth functions can be approximated by polynomials and this approximation guarantees that wavelet coefficients are "sufficiently" small. All this motivates the following definition of *Lipschitz continuity* as a measure of smoothness:

Definition 3.2 *A function f is called (uniformly) Lipschitz α over an interval $[a, b]$ if for all $x \in [a, b]$ there exists a polynomial $p_x(t)$, and there exists a constant K, independent of x, so that*

$$\forall t \in [a, b] : |f(t) - p_x(t)| \leq K|t - x|^\alpha. \tag{3.30}$$

A Lipschitz continuous function can be locally approximated by a polynomial. The effect on the wavelet coefficients of such a function is described by the following theorem, due to Jaffard [103, 130]:

Theorem 3.5 *If a function f is uniformly Lipschitz α over $[a, b]$ and if the wavelet function ψ has p vanishing moments with $p \geq \alpha$, then*

$$\exists C \in \mathbb{R}^+, \forall j \in \mathbb{Z}, \forall i = 1 \dots 2^j : |\langle f, \psi_{j,k} \rangle| \leq C 2^{-j(\alpha + \frac{1}{2})}. \tag{3.31}$$

We now have all the elements to study the asymptotic behavior of the minimum MSE threshold λ^* for piecewise Lipschitz α signals, corrupted by white, stationary and Gaussian noise. We work on the bounded interval $[0, 1]$ and assume that the number of singularities is finite and that the function $f(x)$ is bounded. As in the piecewise polynomial case, we call I_0 the set of coefficients corresponding to basis functions not interfering with one of the singularities and I_1 all the other coefficients. The cardinal numbers of these sets are respectively M_0 and M_1.

We call v^* the critical untouched coefficient value, corresponding to the minimum MSE threshold λ^*: for noise-free coefficients below this value, λ^* is too small, for coefficients with larger magnitude, the threshold is too large. The minimum MSE threshold is the best compromise between these two groups and in Section 3.6.1, Equation 3.23 we found a lower bound for the critical coefficient value: $v^* \geq \sigma^2/\lambda^* \geq \sigma^2/2\lambda^*$. This means that if we call

$$F_0 = \{i = 1, \dots, N \mid |v_i| \leq v^*\},$$

and

$$F_L = \{i = 1, \dots, N : |v_i| \leq \sigma^2/2\lambda^*\},$$

then we know that

$$F_L \subset F_0.$$

We call $K_0 = \#F_0$ the number of coefficients beneath the critical value and $K_1 = N - K_0$ the number of coefficients above this value. It is important to note that

$$v_i \in F_0 \Leftrightarrow \frac{\partial r}{\partial \lambda}(v_i, \lambda^*) \leq 0,$$

and so, we can write the equation

$$ER'(\lambda^*) = 0$$

as

$$-\sum_{i \in F_0} \frac{\partial r}{\partial \lambda}(v_i, \lambda^*) = \sum_{i \in F_1} \frac{\partial r}{\partial \lambda}(v_i, \lambda^*), \qquad (3.32)$$

and both sides in this equation have only positive terms.

We suppose that the coefficients are computed from a direct projection of the continuous signal:

$$v_{j,k} = \sqrt{N}\langle f, \psi_{j,k}\rangle = 2^{J/2}\langle f, \psi_{j,k}\rangle.$$

In practice, these values are approximated by a Fast Wavelet Transform on sample values, or pre-filtered sample values.

To have an idea of the asymptotic behavior of the sums in Equation (3.32), we count the number of terms on the left-hand side:

$$K_0 = \#F_0 \geq \#(F_0 \cap I_0) \geq \#(F_L \cap I_0).$$

The coefficients at the jth resolution level in I_0 satisfy $v_i \leq C2^{J/2}2^{-j(\alpha+1/2)}$. So, if a given resolution level j satisfies

$$C2^{(J-2j\alpha-j)/2} \leq \sigma^2/2\lambda^*,$$

we are sure that all I_0-coefficients at that level are in $F_L \subset F_0$. This condition on j can be worked out as:

$$j \geq \frac{J - \frac{2}{\log 2}\log \sigma^2 + \frac{2}{\log 2}\log \lambda^* + \frac{2}{\log 2}\log 2C}{2\alpha + 1}.$$

In the expression on the right-hand side λ^* is expected to depend on J, but we assume that $\log \lambda^*$ can be neglected with respect to J. If this is not the case, this means that the optimal threshold would increase at least linearly with N. This would not pose any problem to our further analysis, but it is rather unlikely to happen, apart from some pathological cases (a zero signal, for instance). We also drop the constant terms in this right-hand side and we express that $K_1 = N - K_0$ must be smaller than the total number of coefficients at scales j not satisfying this condition *plus* the total number of coefficients in I_1:

$$K_1 \leq M_1 + \sum_{j=0}^{\lfloor J/(2\alpha+1) \rfloor} 2^j$$

$$\approx M_1 + \frac{2^{J/(2\alpha+1)+1} - 1}{2 - 1}$$

$$\sim 2 \cdot 2^{J/(2\alpha+1)}$$

$$= 2N^{1/(2\alpha+1)}.$$

So: $$\frac{K_1}{N} \sim 2N^{1/(2\alpha+1)-1} = 2N^{-2\alpha/(2\alpha+1)}.$$

Taking the logarithm of these asymptotics gives:

$$\log K_1 \sim \frac{\log N}{2\alpha + 1},$$

$$\log K_0 = \log(N - K_1) = \log N + \log\left(1 - \frac{K_1}{N}\right) \sim \log N - \frac{K_1}{N} \sim \log N.$$

Actually, we have started from a lower bound for K_0 to find this behavior. Obviously $\log K_0$ cannot grow faster than $\log N$, since $K_0 \leq N$. On the other hand, the behavior of $\log K_1$ is based on an upper bound. Theoretically, $\log K_1$ may grow slower than $\log N/(2\alpha + 1)$. Following the analysis below, it would turn out that in that case, the minimum MSE threshold would grow (a little) faster. The asymptotic behavior that we will find is a minimal one. Since our primary concern for the next chapter is to demonstrate that the threshold does increase to infinity with N, our result is sufficient for further usage.

We are now ready to fill in both sides of Equation (3.32). For the right-hand side, we assume that λ^* increases slower than v_i, and from Lemma 3.3 it then follows that this side behaves like $K_1 2\lambda^*$. For the left-hand side, we use lower bounds, both for the number of coefficients in F_0 as for their asymptotic behavior. We only consider the coefficients in $F_L \subset F_0$ for which Corollary 3.1 gives a lower bound on the asymptotic behavior.

$$K_0 2\sigma^4 \frac{\phi(\lambda)}{\lambda^2} \sim K_1 2\lambda$$

$$\sigma^4 \frac{\phi(\lambda)}{\lambda^3} \sim \frac{K_1}{N - K_1}$$

$$\sigma^4 \frac{\phi(\lambda)}{\lambda^3} \sim \frac{1}{\frac{N}{K_1} - 1}$$

$$\sigma^4 \frac{\phi(\lambda)}{\lambda^3} \sim \frac{1}{\frac{1}{2}N^{2\alpha/(2\alpha+1)} - 1}$$

$$\sigma^4 \frac{e^{-\lambda^2/2\sigma^2}}{\sqrt{2\pi}\sigma\lambda^3} \sim 2N^{-2\alpha/(2\alpha+1)}.$$

Taking the logarithm on both sides leads to the following theorem:

Theorem 3.6 *If a function f is Lipschitz α on $[0, 1]$, except in a finite number of points and the wavelet analysis has p vanishing moments with $p \geq \alpha$, then the minimum MSE-threshold λ^* for de-noising the corrupted observation*

$$y_i = f(i/N) + \eta_i \qquad i = 1, \ldots, N$$

behaves asymptotically as

$$\lambda^* \sim \sqrt{\frac{2\alpha}{2\alpha + 1}} \sqrt{2 \log N} \sigma, \qquad (3.33)$$

if the number of observations N increases.

The factor

$$\sqrt{\frac{2\alpha}{2\alpha + 1}}$$

comes from the fact that for piecewise smooth functions, coefficients with no inter-action with the singularities are not exactly zero. In our analysis, when estimating the number of coefficients K_1 above the critical value v^*, we even neglected the singularity coefficients, compared to these non-zero coefficients with no singularity interaction: the same behavior would appear (as a lower bound) for signals with no singularity at all.

3.7 Function spaces

3.7.1 Lipschitz regularity

Lipschitz continuity is *in se* a local, point-wise description of regularity. A function which is Lipschitz-1-continuous in x is also continuous in this point, but not neces-sarily differentiable. The notion of uniform Lipschitz regularity extends this regu-larity to an interval. For a Lipschitz constant $\alpha = 1$, for instance, this measurement of regularity is stronger than point-wise continuity and even uniform continuity is a weaker statement.

The uniform Lipschitz idea is based on a minimax principle, which is definitely too restrictive for the type of signals we want to describe. As a matter of fact, the functions that we have in mind typically have *no* uniform behavior, they are *piece-wise* smooth . These functions can be approximated arbitrarily well by uniformly Lipschitz functions in the sense of mean square loss. The sequence of approxima-tions converges, so it is a Cauchy sequence, which means that two elements from the sequence can come arbitrarily close. A Cauchy sequence with all elements be-ing Lipschitz α does not necessarily converge in mean square error to a Lipschitz α function. This is serious shortcoming from the approximation theoretic point of view. The Hölder space of uniformly Lipschitz functions $C^\alpha[0,1]$ is said to be *not complete* with respect to the quadratic distance.

The continuous wavelet transform (CWT) proves to be an excellent instrument to characterise point-wise regularity [130, 126, 19]. Even when a function is only sampled in a countable set of points, a CWT analyses regularity up to the scale known.

3.7.2 Besov spaces

The well known Lebesgue spaces $L_p[0,1]$ ($p \geq 1$), equipped with the corresponding norm

$$\|f\|_{L_p[0,1]} = \left(\int_{[0,1]} |f(x)|^p dx \right)^{\frac{1}{p}}$$

are complete, they allow for singularities, but these spaces provide little smoothness guarantee.

Sobolev spaces $W_p^k[0,1]$ [29] are a sort of complete extensions of $C^k[0,1]$ functions in $L_p[0,1]$. The Sobolev norm is defined as:

$$\|f\|_{W_p^k} = \left(\sum_{n=0}^{k} \|f^{(n)}\|_{L_p[0,1]}^p \right)^p ,$$

For $1 \leq p < \infty$, and

$$\|f\|_{W_\infty^k} = \max_{0 \leq n \leq k} \|f^{(n)}\|_{L_\infty[0,1]}.$$

In this definition $f^{(n)}$ stands for (a weak version of) the nth derivative of f. $W_2^k[0,1]$ spaces are often notated as $H^k[0,1]$.

A further generalization leads to Besov spaces [69, 67, 68, 71]. The definition is quite complicated, and involves a couple of additional concepts:
The r-th difference of a function $f \in L_p[0,1]$ is defined by

$$\Delta_h^{(r)} f(x) = \sum_{k=0}^{r} (-1)^k \binom{r}{k} f(x + kh), \tag{3.34}$$

and call

$$\nu_{r,p}(f;t) = \sup_{h \leq t} \|\Delta_h^{(r)} f\|_{L_p[0,1-rh]} \tag{3.35}$$

the r-th *modulus of smoothness* of f in $L_p[0,1]$. Then for $r > s$, and $1 \leq p, q < \infty$,

$$|f|_{B_{p,q}^\alpha} = \left[\int_{[0,1]} \left(\frac{\nu_{r,p}(f;u)}{u^\alpha} \right)^q \frac{du}{u} \right]^{\frac{1}{q}} \tag{3.36}$$

is the Besov semi-norm of f. A semi-norm may be zero for an essentially non-zero function. To eliminate this unwanted situation, we define the Besov norm of $f \in L_p[0, 1]$ as:

$$\|f\|_{B_{p,q}^\alpha} = \|f\|_{L_p[0,1]} + |f|_{B_{p,q}^\alpha}. \tag{3.37}$$

For $q = \infty$, the semi-norm becomes:

$$|f|_{B_{p,\infty}^\alpha} = \sup_{0<t<1} \frac{\nu_{r,p}(f;t)}{t^\alpha}. \tag{3.38}$$

A Besov space is a set functions with finite Besov norm. To understand what kind of functions belong to a given Besov space, it is interesting to look at the wavelet expansion of these functions. Since all Besov spaces are in $L_p[0, 1]$ and

$$\lim_{j\to\infty} V_j = V_L \oplus \left(\bigoplus_{j=L}^{\infty} W_j \right)$$

is dense in $L_2[0, 1]$, the wavelet expansion

$$f(x) = \sum_{k=1}^{2^L} s_{L,k}\varphi_{L,k}(x) + \sum_{j=L}^{\infty} \sum_{k=1}^{2^j} w_{j,k}\psi_{j,k}(x)$$

converges in $L_p[0, 1]$ sense. Indeed, $L_q[0, 1] \subset L_p[0, 1]$ for $q \geq p$. So, for $p > 2$, we have that $L_p[0, 1] \subset L_2[0, 1]$. Suppose that the mother function $\psi \in L_p[0, 1]$, then all elements of the expansion as well as the limit function are in $L_p[0, 1]$. The error then converges in $L_p[0, 1]$-norm. On the other hand, it is straightforward to see that $L_2[0, 1]$ is a dense subset of $L_p[0, 1]$ with respect to the $L_p[0, 1]$-norm if $1 \leq p < 2$. V_∞ is dense in $L_2[0, 1]$, with respect to the $L_2[0, 1]$-norm, and the Hölder inequality learns that for $p < 2$ (i.e. $2/p > 1$) and q satisfying $p/2 + 1/q = 1$:

$$\left[\int_{[0,1]} |f(u)|^p du \right]^{\frac{1}{p}} \leq \left[\left(\int_{[0,1]} (|f(u)|^p)^{\frac{2}{p}} du \right)^{\frac{p}{2}} \left(\int_{[0,1]} 1^q du \right)^{\frac{1}{q}} \right]^{\frac{1}{p}}$$

$$= \left(\int_{[0,1]} |f(u)|^2 \right)^{\frac{1}{2}}.$$

So all approximations that converge with respect to the $L_2[0, 1]$-norm, do so for the $L_p[0, 1]$-norm as well, provided $1 \leq p \leq 2$. Hence, V_∞ is dense in $L_2[0, 1]$, with respect to the $L_p[0, 1]$-norm. All elements together lead to the conclusion that V_∞ is dense in $L_p[0, 1]$, with respect to the $L_p[0, 1]$-norm: a function in $L_p[0, 1]$ has a converging wavelet expansion.

It turns out that both an upper bound and a lower bound for the Besov norm of a function in $L_p[0, 1]$ can be expressed in terms of this expansion. Call the Besov sequence space:

$$\|w\|_{b^\alpha_{p,q}} = \left[\sum_{j=L}^{\infty} 2^{j\beta q} \left(\sum_{k=1}^{2^j} |w_{j,k}|^p \right)^{\frac{q}{p}} \right]^{\frac{1}{q}}, \qquad (3.39)$$

with $\beta = \alpha + 1/2 - 1/p$. For $q = \infty$, this is:

$$\|w\|_{b^\alpha_{p,\infty}} = \sup_{j \geq L} \left[2^{j\beta} \left(\sum_{k=1}^{2^j} |w_{j,k}|^p \right)^{\frac{1}{p}} \right]. \qquad (3.40)$$

Then there exist constants c and C, not depending on f so that:

$$c\|f\|_{B^\alpha_{p,q}} \leq \|w\|_{b^\alpha_{p,q}} \leq C\|f\|_{B^\alpha_{p,q}}. \qquad (3.41)$$

The wavelet basis is an *unconditional* basis for the Besov space, since the absolute values of the coefficients suffice to check whether a function belongs to the space or not.

This norm equivalence also allows for a characterization of functions in Besov spaces: the wavelet coefficients should decay sufficiently fast to have an expansion with finite Besov sequence norm, and hence a finite Besov function norm. Piecewise smooth functions with a finite number of singularities are among these functions [130]. More thorough interpretations are in [4, 47].

While linear smoothing performs optimally in Sobolev spaces, those methods are sub-optimal in Besov spaces. Functions in Besov spaces typically are piecewise smooth, and here the power of non-linear methods shows up. The statistical optimality properties discussed in the papers by Donoho, Johnstone and collaborators clearly illustrate this observation. One of these results states that the asymptotic behavior of the minimum risk threshold (3.33) for piecewise Lipschitz-*alpha* functions coincides with the minimax risk threshold for functions from Besov spaces $B^\alpha_{p,q}$ [108, 79]. It would be interesting to check whether the minimum risk threshold for functions in Besov spaces has a similar behavior as in Theorem 3.6.

3.7.3 ℓ_p balls

The close connection between Besov spaces and wavelet bases (3.41) illustrates the *multiscale* character of a Besov norm. This becomes clear when looking at expression (3.39). Since thresholding in the first place appeals to *sparsity*, there exists many results on thresholds for sequences with few non-zero elements.

Sparsity can be expressed by looking at the ℓ_p-norm (for $p < 2$) of a sequence $s \in \mathbb{R}^N$:

$$\|s\|_p = \left(\sum_{k=1}^{N} |s_k|^p \right)^{\frac{1}{p}}. \qquad (3.42)$$

One way to control sparsity is imposing a so called *weak ℓ_p-constraint*: if the ordered absolute sequence has sufficiently fast decay

$$|s|_{(k)} \leq \rho N^{\frac{1}{p}} k^{-\frac{1}{p}}, \qquad k = 1, \ldots, N,$$

s is said to belong to a *weak ℓ_p-ball*. Especially when p is small, this favors sequences with lots of zeros. The notion of *strong ℓ_p-balls* follows from a constraint direct on the ℓ_p-norm: a sequence belongs to a (strong) ℓ_p-ball if

$$\frac{1}{N} \sum_{k=1}^{N} |s_k|^p \leq \rho^p,$$

ρ being the 'radius' of the ball.

Optimality results for sequences in ℓ_p-balls include minimax risk analyses [79, 77] and, as mentioned before, the minimax performance of a FDR-procedure [3].

The link between the sparsity in ℓ_p-balls and wavelets is through the concept of Besov sequences. The Besov sequence norm is a weighted, multiscale extension of the ℓ_p-norm. Smoothness in ℓ_p-balls means sparsity, smoothness in Besov spaces is multiscale sparsity. A wavelet analysis is a multiscale, sparse data representation. Some of the results for ℓ_p-balls are the basis for results in Besov spaces [77, 80]. This connection indicates that exploiting the multiresolution structure of wavelet representation may lead to important improvements, both from the theoretical as from the practical point of view. For instance, no global threshold procedure can avoid the logarithmic factor in (3.18), but this factor can be reduced by making the threshold depend on the resolution level. Section 5.1 further discusses the benefits from level dependent thresholding.

3.8 Conclusion

We have proven that the minimum risk threshold is slowly growing if the sample size increases. For piecewise polynomials, the minimum risk threshold asymptotically coincides with the universal threshold, for general piecewise smoothness, the minimum risk threshold is lower, but it comes close to the universal threshold within a constant factor.

We are now ready to motivate an estimation procedure for this minimum risk threshold. The following chapter introduce a generalized cross validation approach and proves that it has asymptotically optimal quality.

4

Estimating the minimum MSE threshold

The previous chapter has investigated the behavior of the minimum risk threshold. In practical problems, the mean square error function can never be evaluated exactly, because the uncorrupted coefficients are necessary to compute the error of the output. Therefore, we need to estimate this MSE function.

This chapter examines a generalized cross validation (GCV) procedure and shows that this leads to an estimate of the MSE-function, the so called GCV-function. So, like MSE, GCV is a function of the threshold value, but evaluation of this function only requires input data and yet its expected value is asymptotically a vertical translation of the risk function. Hence the minimum of this GCV can serve as an estimate for the minimum MSE threshold.

The optimality of GCV is only an asymptotic one. The behavior of GCV for finite data, discussed in Section 4.4, explains why we cannot expect more. To prove this asymptotic properties, we use the knowledge about the asymptotics of the minimum risk threshold from the previous chapter.

GCV is an asymptotically optimal threshold estimator. Speed is a second property: the GCV procedure finds and applies a threshold with less operations than necessary for a wavelet transform. Third, this procedure only uses input data: no additional knowledge or estimations are needed, even not on the amount of noise (the input variance).

Cross Validation is a widely used method for evaluating the optimality of a smoothing parameter. Applications in wavelet based smoothing appear in diverse algorithms [136, 143, 177, 178].

GCV in threshold applications is based on Stein's Unbiased Risk Estimation (SURE). Section 4.1 explains the basics of this threshold estimation method. What follows, is an original investigation of the GCV method. The theoretical argument is illustrated by several test examples.

Like the previous one, this chapter is based on the idea of sparsity in a wavelet representation.

4.1 SURE, a first estimator for the MSE

4.1.1 The effect of the threshold operation

We are looking for an estimator for $R(\lambda)$, which is based on known variables. Therefore we first investigate the effect of the threshold operation on the input data.
    Define

$$F(\lambda) = \frac{\sum_{i=1}^{N}(w_{\lambda i} - w_i)^2}{N} = \frac{1}{N}\|w_\lambda - w\|^2. \tag{4.1}$$

For the expectation of this function, we can write:

$$
\begin{aligned}
\mathrm{E}F(\lambda) &= \mathrm{E}\frac{1}{N}\left(\|w - v\|^2 + \|v - w_\lambda\|^2 + 2\cdot\langle(w - v),(v - w_\lambda)\rangle\right) \\
&= \sigma^2 + \mathrm{E}R(\lambda) - \frac{2}{N}\mathrm{E}\langle\omega, \omega_\lambda\rangle
\end{aligned}
\tag{4.2}
$$

The following lemma leads to an alternative expression for the third term on the right hand side:

Lemma 4.1 *If the density $\phi(\omega_i)$ is Gaussian, then for soft-thresholding:*

$$\mathrm{E}[\omega_i\omega_{\lambda i}] = \sigma^2\mathrm{P}(|w_i| > \lambda). \tag{4.3}$$

Proof:
The Gaussian density function satisfies a first order differential equation:

$$\omega_i\phi(\omega_i) = -\sigma^2\phi'(\omega_i),$$

which allows to write:

$$
\begin{aligned}
\mathrm{E}[\omega_i\omega_{\lambda i}] &= \int_{-\infty}^{\infty}\omega_{\lambda i}\,\omega_i\,\phi(\omega_i)\,d\omega_i \\
&= -\sigma^2\int_{-\infty}^{\infty}\omega_{\lambda i}\,\phi'(\omega_i)\,d\omega_i \\
&= -\sigma^2\omega_{\lambda i}\,\phi(\omega_i)\Big|_{-\infty}^{\infty} + \sigma^2\int_{-\infty}^{\infty}\frac{\partial\omega_{\lambda i}}{\partial\omega_i}\,\phi(\omega_i)\,d\omega_i.
\end{aligned}
$$

Integration by parts is allowed since $\omega_{\lambda i}(\omega_i)$ is a continuous function, at least if we use soft-thresholding.
It is easy to see that:

$$\frac{\partial\omega_{\lambda i}}{\partial\omega_i} = \begin{cases} 0 & \text{if } |w_i| < \lambda, \\ 1 & \text{otherwise,} \end{cases}$$

from which (4.3) follows.   □

This lemma is in fact a special case of more general results by Hudson [101] and Stein [157].

With respect to the third term in (4.2), and for further use, we define

$$\mu_1(\lambda) = \frac{\sum_{i=1}^{N} \mathrm{E}[\omega_i \omega_{\lambda i}]}{N \sigma^2}.$$
(4.4)

The lemma says that:

$$\mu_1(\lambda) = \frac{1}{N} \sum_{i=1}^{N} \mathrm{P}(|w_i| > \lambda).$$
(4.5)

4.1.2 Counting the number of coefficients below the threshold

We now introduce a matrix:

$$D_{ij} = \frac{\partial w_{\lambda i}}{\partial w_j}.$$
(4.6)

Note that if $i \neq j$, then $D_{ij} = 0$. For $i = j$ we have

$$D_{ii} = \begin{cases} 0 & \text{if } |w_i| < \lambda, \\ 1 & \text{otherwise.} \end{cases}$$

Thus, if $\mathrm{Tr}(D)$ is the trace of D, then

$$\mathrm{Tr}(D) = \#\{i \mid w_{\lambda i} \neq 0\} = N - N_0,$$

where

$$N_0 = \#\{i \mid w_{\lambda i} = 0\}$$
(4.7)

Furthermore we consider the Jacobian matrix A with entries

$$A_{ij} = \frac{\partial y_{\lambda i}}{\partial y_j}.$$
(4.8)

Then it is easy to see that

$$A = \tilde{W}^{-1} \cdot D \cdot \tilde{W},$$
(4.9)

where $\tilde{W}$ is the forward wavelet transform matrix. If the transform is orthogonal and $W = \tilde{W}^{-1}$ is the inverse transform matrix, we can write A as the rectangular two-dimensional inverse transform of D:

$$A = WDW^T.$$

Since $\tilde{W}$ is non-singular,

$$\text{Tr}(A) = \text{Tr}(D).$$

With these notations, and since for a Bernoulli variable

$$ED_{ii} = P(D_{ii} = 1), \tag{4.10}$$

we can rewrite μ_1 as:

$$
\begin{aligned}
\mu_1(\lambda) &= \frac{1}{N} \sum_{i=1}^{N} P(D_{ii} = 1) \\
&= \frac{1}{N} \sum_{i=1}^{N} ED_{ii} \\
&= \frac{\text{Tr}(EA)}{N}. \tag{4.11}
\end{aligned}
$$

Starting from $\langle \omega, \omega_\lambda \rangle$, which is not computable in practice, we end up with $\sigma^2 \text{Tr}(A)$, which is easy to find while both have the same expectation. Thus, from (4.2), (4.4), and (4.5) we can construct

$$SURE(\lambda) = F(\lambda) - \sigma^2 + 2\sigma^2 \cdot \frac{\text{Tr}(A)}{N} \tag{4.12}$$

as an approximation for $R(\lambda)$. Application of Stein's Unbiased Risk Estimator [157] leads to the same result [78]. The unbiasedness of this estimator is not an asymptotic property, as it is the case for GCV-optimality (see further). The number of coefficients plays no role, at least not for the expected value.

The estimator can be computed in the original data domain, but since

$$\text{Tr}(A) = \text{Tr}(D) = N - N_0,$$

it is easier to count the number of zero coefficients in the wavelet domain.

4.1.3 SURE is adaptive

Unlike the universal threshold

$$\lambda_{\text{UNIV}} = \sqrt{2 \log N} \sigma,$$

the SURE-threshold does depend directly on the given input signal, not just through a data based estimation of the noise variance σ^2. We may expect a more adaptive threshold choice. It turns out that it is particularly interesting to look for a separate threshold on each scale. So, all coefficients within the same resolution level are put together in one data vector. Donoho and Johnstone explain how a level-dependent minimum SURE procedure then adapts itself automatically to the smoothness class (Besov space) in which the uncorrupted signal probably lies: the method attains asymptotically the minimax behavior within a constant factor, and it does so simultaneously for all spaces taken from a *scale* of Besov spaces [78]. This means that if the uncorrupted function lies in a Besov smoothness space $\mathcal{F} = B_{p,q}^s$, the SURE-algorithm acts as a near-minimax procedure, i.e. it is nearly the estimator that minimizes the worst case risk within the function space:

$$\sup_{f \in \mathcal{F}} \text{Risk}(SURE) \leq C \cdot \inf_{\hat{f}} \sup_{f \in \mathcal{F}} \text{Risk}(\hat{f}), \tag{4.13}$$

where the infimum is taken over all possible estimators $\hat{f}$.

The constant C depends on the smoothness parameters p, q, s of the Besov space, but the result holds simultaneously for all spaces in a scale. This is why the procedure automatically adapts itself to the smoothness class of the uncorrupted signal. Actually, this constant C is not due to the SURE-estimate, but rather to the threshold procedure itself. As a matter of fact, SURE performs asymptotically as well as the minimax threshold on a given Besov space: threshold procedures do not need additional knowledge about the smoothness of the underlying signal to find the (nearly) best threshold.

With respect to this near-minimaxity the performance of SURE is better than that of the universal threshold by a logarithmic factor: compare (4.13) with (3.18). Going level-dependent does not improve universal thresholding for white, homoscedastic input noise and orthogonal wavelets: the universal threshold only depends on the data through an estimation of the noise deviation σ and this deviation is a constant.

4.2 'Ordinary' Cross Validation

This section introduces the idea of Cross Validation in an informal way. Our aim is to minimize the error function based on an unknown exact signal. We therefore try to find a good compromise between goodness of fit and smoothness. We assume that the original signal is *regular* to some extent, which means that the value f_i can be approximated by an linear combination of its neighbors. So, by considering $\tilde{y}_i$, a combination of y_j, not depending on y_i itself, we can eliminate the noise in this particular component. Since we replace it by a weighted average of its neighbors, noise in these components is smoothed, and so we end up with a relatively clean, noise-independent value. This value can be used in the computation of an approximation for $R(\lambda)$.

To investigate the closeness of fit, we compute the result of the threshold operation Λ for the modified signal $\tilde{y}$, in which the i–th component y_i was replaced by $\tilde{y}_i$, i.e.,

$$\tilde{y} = \Lambda \left([y_1, \dots, y_{i-1}, \tilde{y}_i, y_{i+1}, \dots, y_N]^T \right).$$

We then consider the ability of $\tilde{y}_{\lambda i}$ to "predict" the value y_i as a measure for the optimality of the choice of the threshold [56].

For (too) small values of λ the difference $y_i - \tilde{y}_{\lambda i}$ is dominated by noise, while for large values of λ the signal itself is too much deformed. We repeat the same procedure for all components and compute

$$OCV = \frac{1}{N} \sum_{i=1}^{N} (y_i - \tilde{y}_{\lambda i})^2 \qquad (4.14)$$

to express the compromise. This function is called "(Leaving-out-one) Ordinary Cross Validation". This name indicates that we use the values of the *other* components in the calculation for one point. Every function evaluation of (4.14) implies N complete threshold procedures, forward and inverse transform included.

Many combination formulas are possible for $\tilde{y}_i$. Most obvious is to take $\tilde{y}_i = \frac{1}{2} \cdot (y_{i-1} + y_{i+1})$. [135, 136] But taking $\tilde{y}_i$ so that $\tilde{y}_{\lambda i} = \tilde{y}_i$ turns out to be an interesting choice: it leads to an approximating formula for OCV. This value always exists, since the threshold algorithm has a levelling effect. Indeed, taking $\tilde{y}_{\lambda i} = \max_i y_i$, we obtain $\tilde{y}_{\lambda i} \le \tilde{y}_i$, while the opposite is true for $\tilde{y}_{\lambda i} = \min_i y_i$. So, by continuity arguments, one can expect such a value to exist.

For this last choice of $\tilde{y}_i$ we can write:

$$y_i - \tilde{y}_{\lambda i} = \frac{y_i - y_{\lambda i}}{1 - a_i^*},$$

with:

$$a_i^* = \frac{y_{\lambda i} - \tilde{y}_{\lambda i}}{y_i - \tilde{y}_{\lambda i}} = \frac{y_{\lambda i} - \tilde{y}_{\lambda i}}{y_i - \tilde{y}_i} \approx \frac{\partial y_{\lambda i}}{\partial y_i} = A_{ii}.$$

So we have:

$$OCV \approx \frac{1}{N} \sum_{i=1}^{N} \alpha_i^2(\lambda)(y_i - y_{\lambda i})^2,$$

with:

$$\alpha_i(\lambda) = \frac{1}{(1 - A_{ii})}.$$

This expression cannot be evaluated in the wavelet domain and the computation of the matrix A using (4.9) is still cumbersome. We have to minimize the function, and so for each evaluation we need an inverse wavelet transform to compute y_λ as well as the computation of A. Moreover, our experiments indicate that the evaluation is an ill-posed problem, especially for small threshold values, when most of the A_{ii} are close to one. To speed up computations, we can use a kind of an average value for $\alpha_i(\lambda)$:

$$\alpha_i(\lambda) = \alpha(\lambda) = \frac{1}{\frac{1}{N}\sum_{i=1}^{N}(1 - A_{ii})} = \frac{1}{\frac{1}{N}(\mathrm{Tr}(I - A))} = \frac{1}{\frac{1}{N}(\mathrm{Tr}(I - D))}.$$

This gives us the formula of the so called "Generalized Cross Validation" [56, 174, 177, 178]. It turns out that this function can be evaluated and minimized in the wavelet domain. The next section also gives a more mathematical basis for this estimator.

4.3 Generalized Cross Validation

4.3.1 Definition

Generalized Cross Validation is a function of the threshold value:

Definition 4.1

$$GCV(\lambda) = \frac{\frac{1}{N} \cdot \|y - y_\lambda\|^2}{[\frac{\text{Tr}(I-A)}{N}]^2} = \frac{F(\lambda)}{\left(\frac{N_0(\lambda)}{N}\right)^2}, \tag{4.15}$$

with $F(\lambda)$ as in (4.1) and $N_0(\lambda)$ as in (4.7).

With this definition, $GCV(\lambda)$ would become infinite if $N_0 = 0$. For signals in the presence of Gaussian noise this is of course extremely unlikely to happen if $\lambda > 0$, but yet $P(N_0 = 0) \neq 0$, as long as N is finite. This would cause problems for $EGCV(\lambda)$. Therefore we explicitly set $GCV(\lambda) = 0$ if $N_0 = 0$. Another value but zero is of course also possible: since $P(N_0 = 0)$ is close to zero, this has little influence on $EGCV(\lambda)$.

If the wavelet transform is orthogonal, the same formula can be used, mutatis mutandis, in the wavelet domain.

$$GCV(\lambda) = \frac{\frac{1}{N} \cdot \|w - w_\lambda\|^2}{\left(\frac{N_0(\lambda)}{N}\right)^2}.$$

Minimizing this function can be done in the wavelet domain. The denominator is extremely easy to find: just count the number of coefficients below the threshold.

4.3.2 The link between GCV and SURE

Section 4.2 explained how the idea of cross validation can be generalized to the GCV formula. This however gives no guarantee that GCV works fine in concrete applications, and as a matter of fact, it might look surprising how the expression (4.15) leads to an estimation of the MSE function. Next section uses asymptotic arguments to verify under which conditions $GCV(\lambda)$ is a good estimate of $MSE(\lambda)$.

The key to this asymptotical analysis is a relation between GCV and SURE. This link also allows for an intuitive explanation of the working of GCV: if we call $N_1(\lambda) = N - N_0(\lambda)$ the number of coefficients 'surviving' a threshold λ, and if we note that, due to sparseness, N_1/N is typically small, we are able to write:

$$
\begin{aligned}
GCV(\lambda) &= \frac{F(\lambda)}{\left(\frac{N_0(\lambda)}{N}\right)^2} = \frac{F(\lambda)}{\left(1 - \frac{N_1(\lambda)}{N}\right)^2} \\
&\approx \frac{F(\lambda)}{1 - \frac{2N_1(\lambda)}{N}} \approx F(\lambda)\left(1 + \frac{2N_1(\lambda)}{N}\right) \\
&= F(\lambda) + \frac{2N_1(\lambda)}{N}F(\lambda).
\end{aligned}
$$

We know:

$$
\begin{aligned}
F(\lambda) &= EF(\lambda) + [F(\lambda) - EF(\lambda)] \\
&= \sigma^2 + ER(\lambda) - \sigma^2 \frac{2EN_1(\lambda)}{N} + [\text{noisy fluctuations}].
\end{aligned}
$$

Evidently, all four terms in this expression play their role in minimizing $GCV(\lambda)$. For threshold values near to the minimum risk threshold however, $ER(\lambda)$ is much smaller than $\sigma^2 = ER(0)$. After multiplication with $\frac{2N_1(\lambda)}{N}$, the three last terms have little influence when added to $F(\lambda)$. This leads to the approximation:

$$
GCV(\lambda) \approx F(\lambda) + \frac{2N_1(\lambda)}{N}\sigma^2.
$$

Comparison with the definition of SURE (4.12) allows to conclude:

$$
GCV(\lambda) \approx SURE(\lambda) + \sigma^2 \approx MSE(\lambda) + \sigma^2.
$$

Clearly, adding a constant term σ^2 has no influence on the minimum of the objective function. The next section investigates this relation from an asymptotic point of view.

Basically, we assumed that $F(\lambda)$ can be approximated as σ^2 whenever it appears in a term of higher order (*in concreto*: when multiplied with the small factor N_1/N). Putting $F(\lambda) \approx \sigma^2$ means that we suppose that an optimal threshold approximately removes an amount of 'energy' equal to the noise energy. The idea behind this is that a good threshold in a sparse representation removes nearly all the noise, while leaving the information intact. This idea could also serve as an objective when σ is known. Looking for the threshold λ which removes an amount of energy equal to σ^2 corresponds to zeroing the first two terms in the definition (4.12) of $SURE(\lambda)$. Since the third term is small, this is close to minimizing $SURE(\lambda)$, but an exact minimization has of course a better expected result.

4.3.3 Asymptotic behavior

In this paragraph we prove that if $\lambda^* = arg\ min\ R(\lambda)$ and $\hat{\lambda} = arg\ min\ GCV(\lambda)$, then for $N \to \infty$, both minimizers yield a result of the same quality:

$$
\frac{ER(\hat{\lambda})}{ER(\lambda^*)} \to 1. \tag{4.16}
$$

The first difficulty is due to the fact that, unlike in the spline case, or other linear smoothing procedures [9, 8, 92], $GCV(\lambda)$ is a quotient of two variables both depending on the input signal. Next, we compare the result obtained by the minimal GCV-threshold $\hat{\lambda}$ with the result for the optimal threshold λ^*. We give an upper bound for the ratio $\frac{R(\hat{\lambda})}{R(\lambda^*)}$. Finally we show that this upper bound tends to one.

A quotient of two random variables. $GCV(\lambda)$ is a ratio of two random, mutually dependent, variables. We therefore use asymptotic arguments [21] to obtain that for $N \to \infty$:

$$
\begin{aligned}
\frac{ER(\lambda) - (EGCV(\lambda) - \sigma^2)}{ER(\lambda)}
&= 1 - \frac{EGCV(\lambda)}{ER(\lambda)} + \frac{\sigma^2}{ER(\lambda)} \\
&\approx 1 - \frac{ER + \sigma^2 - 2\sigma^2 \mu_1}{(1 - \mu_1)^2 \cdot ER} + \frac{\sigma^2}{ER} \\
&= (1 - \frac{1}{(1 - \mu_1)^2}) + \frac{\sigma^2}{ER} \cdot [\frac{-1 + 2\mu_1}{(1 - \mu_1)^2} + 1] \\
&= \frac{-\mu_1(2 - \mu_1)}{(1 - \mu_1)^2} + \frac{\sigma^2}{ER} \cdot \frac{\mu_1^2}{(1 - \mu_1)^2}.
\end{aligned}
$$

Because $\mu_1 \leq 1$, we have $2\mu_1 \geq \mu_1^2$, and so:

$$
\begin{aligned}
\left| \frac{ER(\lambda) - (EGCV(\lambda) - \sigma^2)}{ER(\lambda)} \right|
&\leq \frac{1}{(1 - \mu_1)^2} \left(|-2\mu_1 + \mu_1^2| + \left| \frac{\sigma^2 \mu_1^2}{ER} \right| \right) \\
&= \frac{1}{(1 - \mu_1)^2} \left(2\mu_1 - \mu_1^2 + \frac{\sigma^2 \mu_1^2}{ER} \right) \\
&\leq \frac{1}{(1 - \mu_1)^2} \left(2\mu_1 + \frac{\sigma^2 \mu_1^2}{ER} \right) =: h(\lambda).
\end{aligned}
$$

If $EGCV(\hat{\lambda}) = \min_\lambda EGCV(\lambda)$ and $ER(\lambda^*) = \min_\lambda ER(\lambda)$, then:

$$
[1 - h(\hat{\lambda})]ER(\hat{\lambda}) \leq EGCV(\hat{\lambda}) - \sigma^2 \leq EGCV(\lambda^*) - \sigma^2 \leq [1 + h(\lambda^*)]ER(\lambda^*),
$$

or:

$$
1 \leq \frac{ER(\hat{\lambda})}{ER(\lambda^*)} \leq \frac{1 + h(\lambda^*)}{1 - h(\hat{\lambda})}. \tag{4.17}
$$

Limit behavior of the upper bound. If $h(\lambda) \to 0$, then $ER(\hat{\lambda}) \to ER(\lambda^*)$. This means that the GCV procedure asymptotically yields a minimum risk threshold. To this end, it is sufficient that $\mu_1(\lambda) \to 0$ and $\frac{\mu_1^2(\lambda)}{ER(\lambda)} \to 0$.

Like in the previous chapter, we first consider the piecewise polynomial case. We call again:

$$
\begin{aligned}
I_0 &= \{i = 1, \ldots, N | v_i = 0\} \\
I_1 &= \{i = 1, \ldots, N | v_i \neq 0\} \\
M_0 &= \#I_0 \\
M_1 &= \#I_1
\end{aligned}
$$

We find for $\mu_1(\lambda^*)$:

$$
\begin{aligned}
\mu_1(\lambda^*) &= \frac{1}{N} \sum_{i=1}^{N} P(|w_i| > \lambda^*) \\
&\leq \frac{\sum_{i \in I_1} 1 + \sum_{i \in I_0} P(|w_i| > \lambda^*)}{N} \\
&\sim \frac{M_1}{N} + \frac{M_0}{N} \frac{1}{\sqrt{\pi N} \sqrt{\log N}},
\end{aligned}
$$

where we used the asymptotic equivalence for the cumulative Gaussian (3.20) and filled in $\lambda^* \sim \sqrt{2 \log N} \sigma$, which indeed tends to infinity.

So, if $N \to \infty$, we know that $\mu_1(\lambda) \to 0$ in the neighborhood of $\lambda = \lambda^*$:

$$
\mu_1(\lambda^*) = \mathcal{O}\left(\frac{\log N}{N} + \frac{1}{N \sqrt{\log N}} \right) \to 0.
$$

To show that $\frac{\mu_1^2(\lambda^*)}{ER(\lambda^*)} \to 0$, we use the fact that for positive a, b, c, d:

$$
\frac{a}{b} < \frac{c}{d} \Rightarrow \frac{a}{b} < \frac{a+c}{b+d} < \frac{c}{d}.
$$

Using the notation $r(v, \lambda)$ from (3.9), we have:

$$
\frac{\sigma^2 \mu_1}{ER} = \frac{\sum_{i=1}^{N} E(\omega_i \omega_{\lambda i})}{\sum_{i=1}^{N} r(v_i, \lambda)} \leq \max_{i=1 \ldots N} \frac{E(\omega_i \omega_{\lambda i})}{r(v_i, \lambda)}.
$$

If $v_i \neq 0$, we have:

$$
\lim_{\lambda \to \infty} \frac{E(\omega_i \omega_{\lambda i})}{r(v_i, \lambda)} = \frac{\lim_{\lambda \to \infty} \sigma^2 P(|w_i| > \lambda)}{\lim_{\lambda \to \infty} r(v_i, \lambda)} = \frac{0}{v_i^2} = 0.
$$

We concentrate on the case $v_i = 0$, and use Lemma 3.2 (expression (3.10) for $v = 0$):

$$
\sigma^2 \frac{P(|w_i| > \lambda)}{r(0, \lambda)} = \frac{2\sigma^2 [1 - \Phi(\lambda)]}{2(\sigma^2 + \lambda^2)[1 - \Phi(\lambda)] - 2\sigma^2 \lambda \phi(\lambda)}.
$$

A long but trivial calculation shows that

$$
\frac{\sigma^2 [1 - \Phi(\lambda)]}{(\sigma^2 + \lambda^2)[1 - \Phi(\lambda)] - \sigma^2 \lambda \phi(\lambda)} \sim \frac{\lambda^2}{2\sigma^2} \qquad \lambda \to \infty. \qquad (4.18)
$$

For the minimum risk threshold $\lambda^* \sim \sqrt{2 \log N} \sigma$, this means that:

$$
\frac{\sigma^2 \mu_1^2(\lambda^*)}{ER(\lambda^*)} = \mu_1(\lambda^*) \frac{\sigma^2 \mu_1(\lambda^*)}{ER(\lambda^*)} = \mathcal{O}\left(\frac{\log^2 N}{N} \right) \to 0.
$$

General piecewise smooth signals. We now turn to the non-polynomial case. Probably none of the uncorrupted coefficients will be exactly zero, so we define:

$$
\begin{aligned}
I_0^\varepsilon &= \{i = 1, \dots, N \,|\, |v_i| \le \varepsilon\lambda^*\} \\
I_1^\varepsilon &= \{i = 1, \dots, N \,|\, |v_i| > \varepsilon\lambda^*\} \\
M_0^\varepsilon &= \#I_0^\varepsilon \\
M_1^\varepsilon &= \#I_1^\varepsilon,
\end{aligned}
$$

where $0 < \varepsilon < 1$ is some arbitrarily small value . We seek an upper bound for M_1^ε. To this end, we determine all levels j where coefficients are certainly below $\varepsilon\lambda^*$ if the corresponding basis function does not interfere with one of the singularities. A similar argument as in Section 3.6.3 leads to:

$$
j \ge \frac{J - \frac{\log J}{\log 2} - \frac{2\log \varepsilon}{\log 2} + \frac{2\log C}{\log 2} - 1 - \frac{\log\log 2}{\log 2}}{2\alpha + 1},
$$

where C and α are the same smoothness parameters as in Section 3.6.3, and J is the finest resolution level ($N = 2^J$). We still follow the same argument as in Section 3.6.3 to find:

$$
\begin{aligned}
M_1^\varepsilon &\le M_1 + \sum_{j=0}^{\lfloor J/(2\alpha+1)\rfloor} 2^j \\
&\sim 2N^{1/(2\alpha+1)}.
\end{aligned}
$$

For the coefficients in the I_0^ε-class, we can write

$$
\mathrm{P}(|w_i| > \lambda^*) \le \mathrm{P}\left(|\omega_i| > (1-\varepsilon)\lambda^*\right),
$$

and so we find, using the asymptotics of Theorem 3.6:

$$
\begin{aligned}
\mu_1(\lambda^*) &\sim \frac{M_1^\varepsilon}{N} + \frac{M_0^\varepsilon}{N} \frac{\sqrt{2\alpha+1}}{\sqrt{2\alpha\pi}(1-\varepsilon)N^{2\alpha(1-\varepsilon)^2/(2\alpha+1)}\sqrt{\log N}} \\
&= \mathcal{O}\left(\frac{1}{N^{2\alpha/(2\alpha+1)}} + \frac{1}{N^{2\alpha(1-\varepsilon)^2/(2\alpha+1)}\sqrt{\log N}}\right) \to 0.
\end{aligned}
$$

And also

$$
\frac{\sigma^2 \mu_1^2(\lambda^*)}{\mathrm{ER}(\lambda^*)} \to 0,
$$

as in the piecewise polynomial case.

Conclusion. In the neighborhood of λ^*, $EGCV(\lambda)$ tends to be a vertical translation of $\mathrm{ER}(\lambda)$ and the relative error of $EGCV(\lambda) - \sigma^2$ has a vanishing upper bound. This leads to the following theorem:

Theorem 4.1 *If* $\mathrm{ER}(\lambda^*) = \min_\lambda \mathrm{ER}(\lambda)$ *and* $\mathrm{EGCV}(\hat\lambda) = \min_\lambda \mathrm{EGCV}(\lambda)$, *then for* $N \to \infty$:

$$\frac{\mathrm{ER}(\hat\lambda)}{\mathrm{ER}(\lambda^*)} \to 1, \tag{4.19}$$

and in the neighborhood of λ^*:

$$\mathrm{EGCV}(\lambda) \approx \mathrm{ER}(\lambda) + \sigma^2. \tag{4.20}$$

Figure 4.1 compares both functions $R(\lambda)$ and $GCV(\lambda)$ for a typical case. The noise variance is 1.1925, the number of data N equals 2048, from which 1984 wavelet coefficients and 64 scaling coefficients, which are not being thresholded.

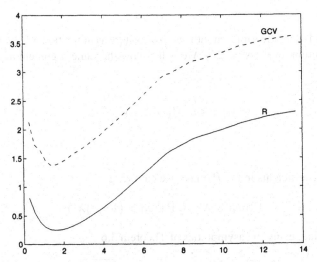

Figure 4.1. GCV and mean square error of the result in function of the threshold λ.

4.4 GCV for a finite number of data

Asymptotic optimality ensures good behavior in most cases, provided that the number of data N is sufficiently large. For signal de-noising applications, our experience is that $N = 1000$ seems to be a minimum for successful application. To illustrate this, we subsample the signal of Figure 3.4, add i.i.d. Gaussian noise ($\sigma = 0.2$) and plot the corresponding GCV and MSE in Figure 4.2. From this example we see that the quality of GCV as an estimator clearly deteriorates for smaller N, although in this case, the minimum GCV threshold remains a reasonable estimate for the minimum MSE-threshold. This is not always the case for small values of N.

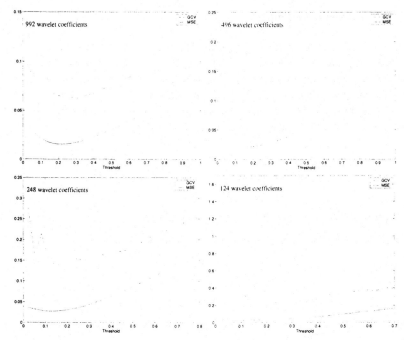

Figure 4.2. GCV and mean square error of the result in function of the threshold λ for different numbers of coefficients.

4.4.1 The minimization procedure

For the moment being, we assume that $GCV(\lambda)$ is a convex function, and we use a Fibonacci minimization procedure [82]. Because $GCV(\lambda)$ is an approximation itself, it is not useful to compute its minimum very precisely. Moreover, in most cases this is not necessary either, due to the smooth curve of $R(\lambda)$ in the neighborhood of its minimum. A relative accuracy of 10^{-4} will do. The Fibonacci procedure attains this error after approximately 20 function evaluations.

Computation of $GCV(\lambda)$ can be performed completely in the wavelet domain. Only at the beginning of the minimization procedure a wavelet transform is needed. As we said before, the denominator in the definition (4.15) counts the number of coefficients that are set to zero. This does not require any floating point operation. Computation of the numerator can be done with $2N$ floating point operations. So 20 function evaluations lead to some $40N$ floating point operations.

For a fast wavelet transform we need $2F2N$ flops, where F is the number of filter coefficients. For $F = 4$, we have $16N$ flops. To reconstruct the signal after the operation with optimal λ, we need an inverse transform too. This makes the minimization procedure not too expensive, as compared with the wavelet transform.

4.4.2 Convexity and continuity

Fibonacci minimization works fine and fast for convex functions. Figure 4.2 however illustrates that $GCV(\lambda)$ is certainly not always strictly convex. As a matter of fact, it is neither continuous: the number of discontinuities almost surely (a.s.) equals the number of data points N. In these points, the right limit is always lower than the left limit, since the denominator increases with one, and between two points of discontinuity the function is a local, increasing parabola:

$$GCV(\lambda) = \frac{N}{N_0^2} \left[\sum_{|w_i|<\lambda} w_i^2 + (N - N_0)\lambda^2 \right].$$

The expected value of this GCV function is continuous:

Theorem 4.2 *If the noise has a non-degenerated Gaussian distribution,* $\mathrm{E}GCV(\lambda)$ *is continuous.*

Proof:

$$\mathrm{E}GCV(\lambda) = \sum_{I \subset \{1,\dots,N\}} \mathrm{E}[GCV(\lambda) \mid |w_i| > \lambda \Leftrightarrow i \in I] \, \mathrm{P}(|w_i| > \lambda \Leftrightarrow i \in I)$$

The second factor in each term is clearly a continuous function of λ, depending of course on the unknown noise-free coefficient values. The first factor equals:

$$\mathrm{E}[GCV(\lambda) \mid |w_i| > \lambda \Leftrightarrow i \in I] = N \frac{\sum_{i \notin I} \mathrm{E}(w_i^2 \mid |w_i| < \lambda) + \#I \cdot \lambda^2}{(N - \#I)^2},$$

which is also continuous   □

   Convexity is not guaranteed, even not in the expected value. Fortunately, the overall curve is "close to convex" so that we do not experience too many problems in minimizing the function.

4.4.3 Behavior for large thresholds and problems near the origin

We examine this function and its singularities at a higher resolution. Figure 4.3 shows a GCV-function, evaluated in 5000 threshold values, instead of 50 as before. We see that most of the discontinuities appear near the origin. This is to be expected, since the major part of the wavelet coefficients are close to zero. Every coefficient value causes a change in the denominator. Actually, this is precisely the mechanism how GCV works: as long as coefficient magnitudes succeed each other at a high rate, many discontinuities appear on a small threshold range. Every discontinuity means a decrease, and the GCV-function has little possibility to 'recover' in between these singularity points. The procedure assumes that this behavior corresponds to the zone of noisy coefficients. The important, large coefficients are far less numerous, and so, from a certain threshold value, the GCV function is able to grow without being 'disturbed' by discontinuities. So, for large threshold values, the function is smoother, and moreover:

$$\lim_{\lambda \to \infty} GCV(\lambda) = \frac{\frac{1}{N}\|0 - w\|^2}{\left(\frac{N}{N}\right)^2} = \frac{1}{N}\|w\|^2.$$

And so:

$$\begin{aligned}
\lim_{\lambda \to \infty} EGCV(\lambda) &= \frac{1}{N}E\|v + \eta\|^2 = \frac{1}{N}\|v\|^2 + \frac{1}{N}E\|\eta\|^2 \\
&= \lim_{\lambda \to \infty} ER(\lambda) + \sigma^2.
\end{aligned} \tag{4.21}$$

For large thresholds and finite N, $EGCV(\lambda)$ behaves like it does asymptotically. This is why we want the minimum risk threshold to increase: the difficulties at the origin persist for $N \to \infty$, while for $\lambda \to \infty$ we get the requested behavior. Since discontinuities happen mainly for typical magnitudes of the numerous noisy coefficients, we may expect success as soon as the minimum risk threshold gets away from the noise level σ. Table 3.1, where $\sigma = 1$, illustrates that this happens for $N \approx 1000$. This corresponds to our experimental findings: for typical signals, we need *grosso modo* thousand samples to guarantee a successful GCV procedure.

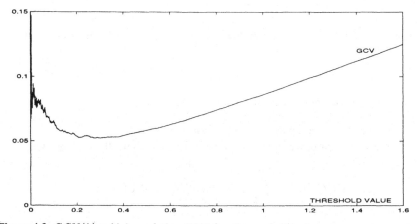

Figure 4.3. $GCV(\lambda)$ at high resolution (5000 function evaluations).

For small values of λ, we have that

$$\lim_{\lambda \to 0} N_0(\lambda) = 0 \qquad a.s.,$$

and in Section 4.3.1 we mentioned that the GCV-definition should treat the case $N_0 = 0$ separately. We could use this opportunity to define

$$GCV(\lambda, N_0 = 0) = 2\sigma^2,$$

so that

$$\lim_{\lambda \to 0} GCV(\lambda) = \sigma^2 + \sigma^2 = \lim_{\lambda \to 0} \mathrm{E}R(0) + \sigma^2 \qquad a.s.,$$

but of course this does not change the difficulties in the neighborhood of the origin, nor has it much influence on the curve of $EGCV(\lambda)$, as the example in the next Section makes clear. Moreover, in practical cases the value of σ is unknown.

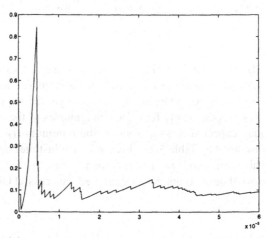

Figure 4.4. $GCV(\lambda)$ for small threshold values. This plot is a detail of the previous one.

4.4.4 GCV in absence of signal and in absence of noise

Two important cases are pure noise and noise-free signals.

In the former case, we can compute $EGCV(\lambda)$ analytically starting from

$$EGCV(\lambda) = \sum_{n=0}^{N} \mathrm{E}\left[GCV(\lambda)|N_0 = n\right] \mathrm{P}(N_0 = n).$$

After some calculations we find:

$$
\begin{aligned}
EGCV(\lambda) \;=\; & 2^N \left[1 - \Phi(\lambda)\right]^N \; GCV(0) \\
& + \sum_{n=1}^{N} \frac{N}{n^2} \binom{N}{n} 2^{N-n} \left[1 - \Phi(\lambda)\right]^{N-n} \left[2\Phi(\lambda) - 1\right]^n \\
& \quad \left[n\sigma^2 \left(1 - \frac{\lambda\phi(\lambda)}{2\Phi(\lambda) - 1}\right) + (N-n)\lambda^2\right]
\end{aligned}
$$

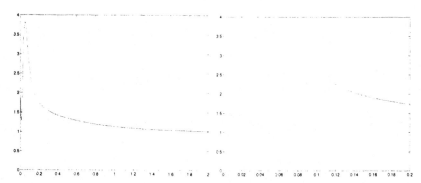

Figure 4.5. $EGCV(\lambda)$ for $f = 0$. The minimum risk threshold is $\lambda = \infty$. This is also the minimum of $EGCV(\lambda)$, if we neglect the minimum near the origin. The right plot is a detail of the left one.

If we choose $GCV(0) = 2\sigma^2$, we get the plot in Figure 4.5 for $\sigma = 1$. We know that $\lambda^* = \infty$ is the minimum risk threshold. It is also a local minimum of $EGCV(\lambda)$, but there is an extra minimum in the immediate neighborhood of the origin.

If the input signal is noise-free, the minimum risk threshold is of course zero. With our experience of difficulties near the origin, we expect a troublesome application of GCV. Of course, the performance depends on the signal characteristics. If the signal looks irregular, almost like noise, GCV cannot detect this as a real signal. But for typical piecewise smooth signals, the procedure does a pretty good job. We have that:

$$EGCV(\lambda) = GCV(\lambda) = \frac{\frac{1}{N}\|w_\lambda - w\|^2}{\left(\frac{N_0}{N}\right)^2} = \frac{\frac{1}{N}\|v_\lambda - v\|^2}{\left(\frac{N_0}{N}\right)^2} = \frac{R(\lambda)}{\left(\frac{N_0}{N}\right)^2}.$$

The problems occur in the region where many coefficient values succeed each other. If the coefficients are affected by noise, this interval stretches from the origin to σ say. But if the signal has no noise, $\sigma = 0$, most of the coefficients are much closer to zero. Even if GCV would fail on this tiny interval, this would hardly affect the result. Figure 4.6 plots $GCV(\lambda)$ for the wavelet coefficients of the test signal in Figure 3.4.

4.4.5 Absolute and relative error

We know that

$$\left| \frac{ER(\lambda) - (EGCV(\lambda) - \sigma^2)}{ER(\lambda)} \right| \leq h(\lambda).$$

This is an upper bound on the relative error of $EGCV(\lambda) - \sigma^2$ as an approximation of $R(\lambda)$. The question arises whether it would not be easier to start from the asymptotic behavior of the minimum risk threshold $\lambda^* \sim \sqrt{2\log N}\sigma$ and to use the fact

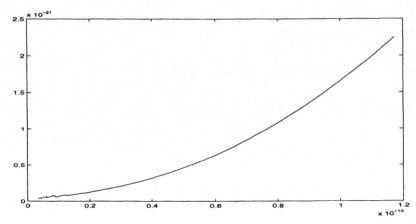

Figure 4.6. $GCV(\lambda)$ for $\eta = 0, \sigma = 0$. Pay attention to the abscis values: even for extremely small threshold values, GCV performs almost perfectly.

that for increasing threshold values, the *absolute* approximation error (4.21) tends to zero.

At least two reasons make this a bad idea. The first is of course that increasing N not only causes λ^* to grow, but also adds more data, and so changes the entire GCV-curve. It is not clear how to use a asymptotic result for a fixed curve in a situation of a simultaneous change of curve and abscis point.

Second, both $EGCV(\lambda^*) - \sigma^2$ and $R(\lambda^*)$ tend to zero for large N. Proving that the absolute difference between these two vanishes, gives little information on the real quality of the GCV-procedure as an estimate.

4.4.6 Which is better: GCV or universal?

We have proven that the quality of the minimum GCV-threshold $\hat{\lambda}$ tends to be optimal for large data sets:

$$\frac{ER(\hat{\lambda})}{ER(\lambda^*)} \to 1.$$

Strictly spoken, this gives no certainty about the asymptotics of $\hat{\lambda}$ itself, and as a matter of fact, we do not need to know how it behaves: if we can count on its *quality* in terms of risk, we have all we want.

Nevertheless, we may *expect* that $\hat{\lambda}$ has the same behavior as λ^*. In the piecewise polynomial case, this coincides with the universal threshold, in the piecewise smooth case, it coincides up to a constant.

This does not mean that the GCV-threshold has the same properties as the universal threshold. For signals with finite length, we may expect that the GCV-procedure, like SURE, is more adaptive to the signal than the universal threshold:

GCV uses all data, and not just through an estimate of the noise level σ. (GCV does not need a value at all for σ.)

Moreover, a similar asymptotic behavior is no guarantee for a similar asymptotic quality. The following — hypothetic — example illustrates this. Suppose the risk is given by the expression:

$$ER(\lambda) = (\lambda - \sqrt{2\log N} + 1/N)^2 + 1/N^3,$$

so we see that the minimum risk threshold would be $\lambda^* = \sqrt{2\log N} - 1/N$. And assume that GCV finds $\hat{\lambda} = \sqrt{2\log N} - 1/N + 1/N^2$ and the noise variance equals 1, so the universal threshold is simply $\sqrt{2\log N}$. These three thresholds all have the same asymptotic behavior. We see that, like in Theorem 4.1

$$\frac{ER(\hat{\lambda})}{ER(\lambda^*)} = \frac{1}{N} + 1 \to 1,$$

but the universal threshold does not attain this quality, in terms of minimum risk:

$$\frac{ER(\lambda_{\text{UNIV}})}{ER(\lambda^*)} = \frac{(1/N)^2 + (1/N)^3}{(1/N)^3} = N + 1 \to \infty.$$

The GCV threshold, being an estimator for the minimum risk threshold, of course shows the same disadvantages. In particular, it tends to leave too much noisy coefficients at finer scales, which causes unwanted 'blips' in the output. The next chapter explains where this problem comes from and gives several possible remedies.

4.5 Concluding remarks

This chapter has introduced and motivated the use of a *generalized cross validation* procedure for finding a good threshold for wavelet coefficients. This method shows the following properties:

1. Minimizing the GCV-function is a fast method: it is not a bottleneck in a wavelet threshold procedure.
2. The method needs no explicit estimation for the noise level σ (or its variance σ^2). This advantage becomes particularly useful when the amount of noise depends on the resolution level. Section 5.1.1 illustrates that this is the case if the input noise is correlated.
3. The method is asymptotically optimal.
4. For finite data, the GCV function shows a difficult behavior near the origin, but this is a unimportant region: it contains threshold values below the noise level σ.

To prove the asymptotic optimality, we had to make several assumptions. Our experiments show that most of these conditions are crucial: if the noise or the wavelet transform does not satisfy one of them, the algorithm does not work as expected:

1. The untouched signal should be *smooth* in the sense that it can be represented sparsely by taking a wavelet transform. In fact, this assumption justifies the use of wavelets, since the decorrelating properties of wavelet transform guarantee such a sparse representation for most noise-free signals and images.
2. The noise in wavelet domain should be homoscedastic. As the following chapter explains, this is because GCV is based on SURE, and we need one σ for a successful application of SURE. To this end, we need:
   a) the input noise to be second order stationary,
   b) the input noise to be uncorrelated,
   c) the wavelet transform to be orthogonal.

   The next chapter goes deeper into this condition and also proposes a relaxation for the case of correlated noise. It also discusses possible schemes to deal with Poisson noise, which is highly non-stationary.
3. The noise should be Gaussian with zero mean. Experiments showed that, in practice, the GCV-method performs well for other zero mean homoscedastic distributions of the noise, provided the density function has a finite variance. A Cauchy density, with its heavy tail, is a typical example of a density for which our estimator is not appropriate.
4. The algorithm should use *soft*-thresholding. Minimizing GCV, while using hard-thresholding for the output mostly leads to a threshold equal to zero.

We also remark that the ideas of this chapter are not strictly limited to *wavelet* thresholding. The GCV procedure is based on the *sparsity* of this representation, and the proofs can probably easily be adapted to other kinds of sparse representations. This chapter shows that the idea of generalized cross validation, well-known in the framework of linear smoothing (like splines) is also applicable to non-linear methods.

5

Thresholding and GCV applicability in more realistic situations

We have exploited the sparsity of a wavelet representation to motivate thresholding as a curve fitting method and to find GCV as an asymptotically optimal threshold assessment procedure. We explained that sparsity is a sort of smoothness and explained that wavelets are well suited to measure this concept of piecewise smoothness. We have seen that smooth reconstructions, e.g. using the universal threshold, show bias. In a context of image processing, bias is blur. On the other hand, the minimum risk threshold often leads to an output with many spurious structures, 'blips', which are remaining noise components.

Until now, we hardly used this other important wavelet characteristic, which is the natural way wavelets support the idea of multiresolution . A wavelet decomposition is not only a sparse representation, it is also a multiscale data representation.

Most of this chapter emphasizes this aspect of a wavelet decomposition. Multiresolution and sparsity together create more possibilities. First, signals mostly have different characteristics at different scales. Indeed, scale can be seen as the approximate inverse of frequency. Just as in signal or image processing lots of operations are based on frequency analysis, we can use the multiscale character of the wavelet transform to make operations more adaptive. Not only should the operations be scale-dependent, but it is also useful to look across scales and handle one scale taking into account the information at adjacent resolutions.

Second, if the noise is correlated instead of white, also the noise behavior depends on the resolution. Clearly, one threshold cannot remove noise decently if the amount of noise σ depends on the resolution level. Scale-dependent thresholds are a solution for correlated noise, and are more adaptive to signal characteristics.

Other modifications to the classical settings are less related to multiresolution. The non-decimated wavelet transform causes additional smoothing, at the price of an $N \log N$ algorithm instead of a linear complexity. The integer wavelet transform avoids the use of floating point numbers, thereby speeding up computations, and eliminating rounding errors.

The modifications in this and next chapter aim at both a visual improvement (more smoothness) and a reduction of the output MSE. The minimum MSE *threshold*, as discussed in the previous chapters, is based on a global compromise between noise and data: this is not the best thing we can do: instead of applying one threshold for all coefficients at a given level, we would like to decide for each coefficient separately what is best: keeping or killing. If we know the noise deviation σ and if an "oracle" would tell us the noise-free magnitude of each coefficient, we could apply the best possible selection from Section 3.4.1, i.e. the minimum risk selection: keep coefficients with uncorrupted value above σ and replace the others by zero. We remind that this is not thresholding, because the decisions are based on the uncorrupted values, not on the noisy ones. Thresholding is one particular example of this general idea of coefficient selection and the minimum risk selection is another one. This oracle selection remains an ideal benchmark, but we hope that using the multiscale nature of a wavelet representation helps in mimicking this oracle.

5.1 Scale dependent thresholding

5.1.1 Correlated noise

To generate correlated or colored noise, we apply a FIR-highpass-filter to white noise. A FIR or finite impulse response filter has a finite number of taps (filter coefficients), for the example of Figure 5.1 we convolve with 100 coefficients. If we add this noise to the test signal in Figure 3.4, we get a picture on top of Figure 5.1 which at first sight does not show any difference from the classical setting. A plot of the GCV and MSE function in Figure 5.2 however indicates that GCV is not able to find an approximation for the minimum risk threshold. The reason for this becomes clear if we plot the wavelet coefficients of the noise in Figure 5.1, middle. This noise is clearly not stationary, and therefore we have a coefficient-depending σ in Lemma 4.1 (4.3). As a consequence, a SURE-formula as in (4.12) is no longer valid. Since the GCV-asymptotic properties are based on this unbiased estimator, we cannot guarantee a successful application of GCV anymore.

From the analysis of the correlation matrix of the noisy wavelet coefficients, we observed that uncorrelated, stationary noise remains stationary after an orthogonal transform. If the noise is not stationary or not white, the wavelet transform could be neither white nor stationary.

To prove that GCV yields the optimal threshold if the number of wavelet coefficients tends to infinity, we do not need uncorrelated wavelet coefficients at any moment, but for the motivation of the SURE-threshold, we do need stationary noise in the wavelet domain.

Even if we would find the minimum MSE threshold, it would not be that useful. Indeed, intuition says that the more the coefficients are affected by noise, the higher the threshold should be. The universal threshold states this explicitly: $\lambda_{UNIV} \propto \sigma$, but also the minimum risk threshold is approximately proportional to the amount of noise. If the amount of noise depends on the coefficient, it is hard to remove it by

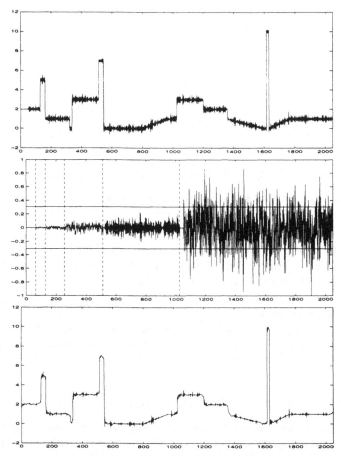

Figure 5.1. A signal with correlated, stationary noise ($\sigma = 0.2$). This noise was generated by convolution of white noise with a FIR-highpass-filter. In the middle: the wavelet transform of the noise. The dashed line indicates the boundaries between successive resolution levels. The two horizontal lines are at $\pm\lambda_{\mathrm{MSE}}$. One threshold, even the minimum MSE threshold, cannot remove noise with different σ simultaneously and decently. The bottom plot shows the reconstruction after applying the minimum MSE threshold: lots of noisy structures from the finest scale have survived the noise reduction.

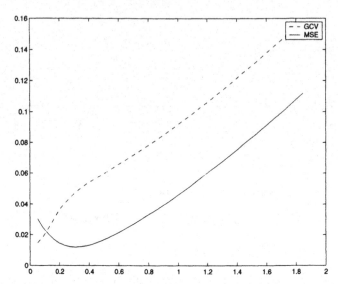

Figure 5.2. *GCV* and mean square error for the signal with correlated noise in Figure 5.1. GCV fails as an estimator of the optimal threshold.

only one threshold. The reconstruction on the bottom of Figure 5.1 comes from a minimum MSE threshold. This threshold is clearly too small to remove the noise at the finest scale: most of the noise is concentrated at this resolution. This is not surprising, since the noise was generated by a highpass-filter and high frequencies correspond to fine scales.

If we happen to know the covariance structure of the input noise, i.e. if we know the covariance matrix up to a constant factor, then we can compute the covariance structure S in wavelet domain, using (2.20), and normalize the wavelet coefficients as:

$$w_i^{\text{norm}} = w_i / \sqrt{S_{ii}}, \qquad (5.1)$$

where, as usual, index i stands for j, k, j indicating scale and k indicating location. Unfortunately, normaliter we do not know the correlation structure of the input, and therefore we count upon the multiresolution character of wavelet transform.

We suppose that the original noise is stationary and more precisely that the correlation between two points only depends on the distance between them. This means that the correlation matrix Q is a (symmetric) Toeplitz matrix. If this is true, the multiresolution structure of a wavelet transform allows to prove that:

Lemma 5.1 *If $\omega_{j,k}$ represents a wavelet coefficient of a stationary random vector η at location k, and resolution level j (scale 2^{-j}), then the variance of this coefficient, $\mathrm{E}(\omega_{j,k})^2$, only depends on the resolution level j.*

This lemma explains why we denote the noise deviation at level j as σ_j.
Proof:
Since the correlation matrix is symmetric Toeplitz, we have that $Q_{u,v} = q_{|u-v|}$.

The wavelet coefficients at the finest resolution level $J - 1$ (where $N = 2^J$) are then:

$$\begin{aligned}
\mathrm{E}\omega_{J-1,k}\omega_{J-1,l} &= \sum_u \sum_v \tilde{g}_{u-2k}\tilde{g}_{v-2l}\mathrm{E}\eta_u\eta_v \\
&= \sum_u \sum_v \tilde{g}_{u-2k}\tilde{g}_{v-2l}q_{|u-v|}.
\end{aligned}$$

Substitutions $m = u - 2k$ and $n = v - 2l$ then yield that:

$$\mathrm{E}\omega_{J-1,k}\omega_{J-1,l} = \sum_m \sum_n \tilde{g}_m\tilde{g}_n q_{|2(k-l)+m-n|}.$$

From this formula, it follows immediately that for all integer r:

$$\mathrm{E}\omega_{J-1,k+r}\omega_{J-1,l+r} = \mathrm{E}\omega_{J-1,k}\omega_{J-1,l}.$$

In particular, we have that:

$$\mathrm{E}\omega^2_{J-1,k+r} = \mathrm{E}\omega^2_{J-1,k} =: \sigma^2_{J-1}.$$

A similar argument holds for the scaling coefficients at resolution level $J - 1$. We can thus repeat the same procedure for the wavelet coefficients at coarser levels, thereby completing the proof.   □

For a two-dimensional wavelet transform, the noise level also depends on the orientation (vertical, horizontal, diagonal) of the coefficient [105].

5.1.2 Level dependent threshold estimation by GCV

We have proven that the wavelet transform of stationary correlated noise is stationary within each resolution level. Since stationarity is a condition for a successful GCV-estimation of the optimal threshold, this result suggests choosing a different threshold for each resolution level.

The mean square error now becomes a function of a vector of thresholds $\boldsymbol{\lambda}$. If $\boldsymbol{w}_j^c$ denotes the vector of wavelet coefficients at resolution level j, and component (orientation) c, then we can write:

$$R(\boldsymbol{\lambda}) = \sum_{j,c} \frac{N_j^c}{N} R_j^c(\lambda_j^c), \tag{5.2}$$

where N_j^c represents the number of wavelet coefficients on level j and component c, and

$$R_j^c(\lambda_j^c) = \frac{1}{N_j^c}\|\boldsymbol{w}_{j,\lambda}^c - \boldsymbol{v}_j^c\|^2. \tag{5.3}$$

Since all terms in (5.2) are positive, minimization of $R(\boldsymbol{\lambda})$ is equivalent to successive one dimensional minimizations of $R_j^c(\lambda_j^c)$ for all j and c. A similar argument as in Chapter 4 leads to an estimation [109]:

$$\text{SURE}_j^c(\lambda_j^c) = F_j^c(\lambda_j^c) + 2(\sigma_j^c)^2 \cdot \frac{N_j^c - N_{j0}^c}{N_j^c}, \tag{5.4}$$

with:

$$F_j^c(\lambda_j^c) = \frac{1}{N_j^c} \| w_{j,\lambda}^c - w_j^c \|^2,$$

and this leads to an adaptive estimation of the underlying signal, just like in the white noise case.

Based on this estimator, we can construct:

$$GCV_j^c(\lambda_j^c) = \frac{\frac{1}{N_j^c} \| w_j^c - w_{j,\lambda}^c \|^2}{\left(\frac{N_{j0}^c}{N_j^c} \right)^2}, \tag{5.5}$$

and the minimizer of this function is an asymptotically optimal estimator for the minimum risk threshold for level j and component c.

In some sense, it is remarkable that a GCV procedure still works for correlated noise. Indeed, it is widely known that the criterion fails in finding a good smoothing parameter in spline smoothing [176] or an optimal bandwidth in kernel estimation [7, 98, 99]. GCV is reported to underestimate the smoothing parameter, thereby yielding an output which is close to a mere interpolation of the noisy input. Basically, we observe the same phenomenon in the plot of Figure 5.2, when trying to find one global threshold. Figure 5.18 shows two examples of level-dependent threshold selection for data with correlated noise. Apparently, and a bit surprisingly, no specific problems occur from the data interdependence.

The reason for this straightforward application of the GCV procedure to data with correlated noise lies in the fact that the properties of GCV are not corrupted by correlated noise directly. Correlated noise only affects the algorithm through the heteroscedasticity of the wavelet coefficients. Fortunately, the multiresolution structure of a wavelet transform assures the noise to remain homoscedastic within one scale. The situation is different for some properties of the *universal* threshold. The 'probabilistic upper bound' from Section 3.4.5, for instance, is no longer preserved in the case of correlated data. This is a reason for using an updated expression in the case of correlated noise [24]. On the other hand, level-dependent universal thresholding with $\lambda_j = \sqrt{2 \log N} \sigma_j$ remains optimal w.r.t. the oracle risk [109].

The fact that GCV needs no explicit estimation for σ_j^c becomes particularly advantageous here: in the white noise case, the noise level is equal at all scales, and it can easily be estimated from the coefficients at the finest scale, since this resolution level generally contains little important coefficients. If the noise is colored however, or the transform is non-orthogonal, we need an estimate at each scale and for each component (in a 2D wavelet transform). At coarser scales, this may be a problem, since relatively many large coefficients with information are present here. On the other hand, the same phenomenon also deteriorates the quality of the GCV-estimation, but we believe that the one-stage GCV-estimation better resists than a procedure in which noise estimation and noise reduction are separated. This is confirmed by a threshold comparison in Table 5.3. At coarse scales, the SURE thresholds suffer from a bad estimation of the noise deviation.

5.1.3 Non-orthogonal transforms

If the input noise is uncorrelated and stationary, but we use a biorthogonal wavelet transform, Equation (2.20) learns that the wavelet coefficients are correlated and not stationary. In this case, we have all the information to normalize the coefficients like in (5.1), but we can also apply level-dependent thresholds. Level dependency has the advantage of signal-adaptivity as we explain in the following section and it still works if the noise is colored.

Another problem rises from the fact that non-orthogonal transforms do not preserve ℓ_2-norms. Stricto sensu, we cannot minimize MSE or GCV in the wavelet domain. Riesz bounds however guarantee a quasi-equivalence of norms. Moreover, as explained in Section 3.1, there seems to be no reason why pixel-MSE corresponds better to visual quality than multiscale-MSE.

5.1.4 Scale-adaptivity

Even for white noise and orthogonal transform, level-dependent thresholding may be interesting. Indeed, the optimal threshold not only depends on the noise level, but of course also on the signal characteristics. These characteristics may differ at different levels. Typically, coarse levels show a larger proportion of important signal features. The presence of large coefficients forces the optimal threshold to smaller values: the curves of the one-coefficient-risk in Figure 3.2 illustrate that large uncorrupted values prefer small thresholds. Although the noise level may be a constant across scales, minimum risk thresholds at coarser scales tend to be smaller. If the algorithm seeks one global threshold, this has to be a trade-off: for the finest scale, it is probably too small, and this shows up in the output as noisy 'blips'. Scale-dependent thresholding is a way of reducing these spurious structures.

This adaptivity advantage also shows up in a theoretical analysis. As explained in Section 3.7.3, no global threshold outperforms the adaptivity of the universal threshold in (3.18). Level dependent thresholding, using SURE thresholds, however avoids the logarithmic term, as explained in Section 4.1.3, formula (4.13). The same result holds for FDR thresholds.

5.2 Non-decimated wavelet transforms

The discussion in Section 2.3 and Figure 2.14 show that the non-decimated wavelet transform has order of complexity $\mathcal{O}(N \log N)$, both for memory as for computations. This is a factor $\log N$ more expensive than the fast wavelet transform.

On the other hand, this redundant transform has some advantages, especially for noise reduction:

1. The non-decimated wavelet transform generates an equal number of coefficients at all resolution levels. In principle, this facilitates the use of an asymptotic method like GCV at coarse scales. The proportion of noise-free coefficients however remains the same: at coarse scales, these are quite numerous, and so, there is not really a sparse representation here, coefficients are highly correlated. This effect partially undoes the benefit from the extra coefficients. Therefore, we still leave the coarsest scales untouched.

2. It is easy to prove that a redundant wavelet transform of stationary noise is still stationary within each scale.

3. This redundant transform is immediately extensible for cases where the number of data is not a power of two.

4. Unlike the decimated transform, this redundant transform is translation invariant. As a matter of fact, the non-decimated wavelet transform is an interleaving rearrangement of all fast wavelet transforms of shifted versions of the input. More precisely, let the input y contain N data points and define the shift operator S as:

$$z = Sy \Leftrightarrow \begin{cases} z_i &= y_{i-1} \text{ for } i = 2, \ldots, N \\ z_1 &= y_N. \end{cases} \tag{5.6}$$

Then the redundant wavelet transform contains all coefficients from

$$w_n = \tilde{W} S^n y, \qquad n = 1, \ldots N, \tag{5.7}$$

where $\tilde{W}$ is the non-redundant transform matrix. In principle, there are $N \times N = N^2$ of these coefficients, but the redundant transform eliminates doubles and rearranges the $N \log N$ remaining coefficients.

5. In each step of the inverse transform, we could omit one half of the (wavelet and scaling) coefficients before reconstruction of the scaling coefficients at the previous level. This means that these coefficients can be reconstructed in two independent ways. If we manipulate the wavelet coefficients, for instance to remove noise, then the result will probably not be an exact redundant wavelet transform of any signal at all. As a consequence the two possible reconstruction schemes at each level generate two different scaling coefficients at the previous level. Experiments show that taking a linear combination of these two possibilities causes an extra smoothing.

It is not hard to understand that taking the simple mean of the two reconstruction schemes in each step corresponds to averaging the reconstructions from all fast wavelet transforms of shifted versions of the input. This is:

$$y_{\text{recon}} = \frac{1}{N} \sum_{n=1}^{N} S^{-n} W w_n, \tag{5.8}$$

where $W = \tilde{W}^{-1}$ is the inverse wavelet transform and w_n has been defined in (5.7).

For manipulated w_n, all terms in (5.8) are different, and averaging causes smoothing. In the case of orthogonal transforms $W = \tilde{W}^T$, y_{recon} is the least squares solution to the overdetermined problem:

$$\tilde{W}_{\mathrm{redun}} y = w_{\mathrm{redun}}.$$

Simple Matlab^{TM} testing shows that this least square interpretation does not hold for biorthogonal transforms. Anyway, the reconstruction from thresholded non-decimated coefficients is smoother, as experiments in Section 5.6 illustrate. Similar observations, with basically the same inverse transform, hold for other threshold as well [53, 137].

5.3 Interscale dependencies

Section 5.1 introduced scale dependent thresholding as a technique adaptive to specific signal or noise characteristics at different scales. This section proposes two algorithms to take into account interscale dependencies. Coefficients at neighboring levels provide additional information about the importance of a given coefficient. Incorporating this information into a coefficient selection scheme leads to a more global approach.

5.3.1 Tree-structured thresholding

If $N \to \infty$, we know that the minimum risk threshold behaves (approximately) as $\lambda^* \sim \sqrt{2 \log N} \sigma$. For general piecewise smooth functions, we are not sure that λ^* increases that fast, but the correction term $\sqrt{2\alpha/(2\alpha + 1)}$ acts as a sort of 'asymptotic lower bound': the proof of Theorem 3.6 shows that λ^* may increase faster. So, for the moment we assume that the minimum risk threshold behaves asymptotically like the universal threshold, which means that for $N \to \infty$ a pure noise coefficients has no chance of passing the threshold.

For finite N, however, this probability is positive, even for the universal threshold, and the minimum risk threshold is still smaller. Level-dependent thresholding does not change this, and so there is always a proportion of noisy coefficients surviving the threshold.

Another problem for level-dependent thresholding comes from the fact that the GCV procedure needs sufficiently many coefficients to work well. Coarser resolution levels may lack this number of coefficients to find a separate threshold.

To further reduce these spurious output structures, we therefore return to one threshold, but appeal to another heuristic about wavelet transforms: if a coefficient at a given scale and location has a large value because it carries signal information, we may expect that the coefficient at the same location and coarser scale also has a large magnitude. This is because signal features typically have a wide range: a signal singularity therefore causes important coefficients at different scales. Noise, on the other hand, is a local phenomenon: a noise singularity does not show up at different scales.

This idea has been used in different alternatives for the classical thresholding. The algorithm by Xu, Weaver, Healy, and Lu [180] selects coefficients on a basis of interscale coefficient correlation instead of simple magnitudes. Other methods [20, 73] select *trees* of wavelet coefficients. A *tree* is a set of wavelet coefficients so that for each element, except one (called the *root*), the coefficient at the same location and at the next, coarser scale also belongs to the tree. Since two different fine scale coefficient share one single 'parent' coefficient, the multiscale representation of this set has a branched structure [20], hence it is called a tree.

Just as for minimum risk thresholding, the optimal tree is the best trade-off between sparsity and closeness of fit. To estimate this 'best tree' w_T, the procedure minimizes a 'complexity-penalized residual sum of squares':

$$CPRESS(w_T) = \|w_T - w\|^2 + \lambda^2(N - N_0),$$

where, as usual, N_0 indicates the number of coefficients in w_T that are exactly zero. In this expression, λ is a smoothing parameter that can be tuned to find a good compromise between smoothness and closeness of fit. When minimizing $CPRESS(w_T)$, we impose two constraints:

1. *Keep or kill*: every coefficient $w_{T,i}$ equals w_i or zero.
2. *Tree*: If a coefficient $w_{T,i} = 0$, then all coefficients at finer scales at the same location must be zero. So, we get a zero-subtree.

Actually, we are minimizing over a binary *label* vector $x \in \{0,1\}^N$:

$$CPRESS(x) = \sum_{i=1}^{N}(w_i x_i - w_i)^2 - \lambda^2 x_i \qquad (5.9)$$

under the constraint that $\{x_i | x_i = 1\}$ must be a tree. This minimization problem can be solved in $\mathcal{O}(N)$ computations, using a dynamic programming algorithm [73, 28]. If we do not impose this tree structure, minimizing (5.9) would lead to a simple hard-thresholding, with threshold λ. Donoho proposes to choose he smoothing parameter as $\lambda \approx \sqrt{2 \log N}\sigma$.

The form of (5.9) is less general than the objective function that leads to hard-thresholding in Chapter 2. To our knowledge, there exists no immediate alternative for (5.9), leading to a sort of soft-thresholding, and allowing for a fast procedure. An algorithm may keep or kill coefficients in a tree, regardless of their magnitude, but it is impossible to shrink coefficients in a tree with a certain value, without actually killing some of them, if there is no *a priori* lower bound on the magnitude of the coefficients in the tree. Since the GCV-procedure is based on the idea of a *continuous* operation like soft-thresholding, it appears to be difficult to incorporate a GCV choice of the smoothing parameter in this tree-structured algorithm.

Nevertheless, we can use the idea that noise is local and only causes accidental values with no correlation across scales. After the threshold operation, we are suspicious of surviving coefficients at fine scales and we check whether the corresponding coefficient at coarser scales also have passed the threshold. Section 5.6 compares this method with others.

5.3.2 An example

We now compare the methods discussed so far. We sample the signal from in $N = 2048$ equidistant points. We add white noise in a signal-to-noise ratio of 10 decibels. This leads to the noisy signal in Figure 5.3. Figure 5.4 compares the

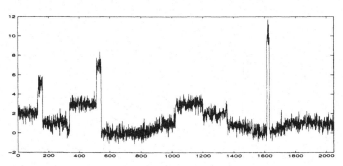

Figure 5.3. Noisy test signal. SNR = 10 dB.

output from different algorithms, all using Daubechies' orthogonal wavelets with 3 vanishing moments. Table 5.1 has the corresponding output SNR-values. Four levels are processed, all other, coarser scale coefficients are left untouched. Table

simple GCV	tree-structured	level-dependent	non-decimated, level-dependent
16.81 dB	17.10 dB	17.06 dB	18.12 dB

Table 5.1. Output SNR-values for different methods of Figure 5.4.

and figure illustrate that signal-to-noise ratio and visual quality do not always coincide. The tree-structured method succeeds best in removing unwanted blips, but the redundant transform achieves a better SNR-value. The next figures contain the GCV-plots. Figure 5.5 shows the global threshold selection, used in the first two outputs. Figure 5.6 shows the four GCV-plots used in the level-dependent threshold algorithm. Even at the coarsest scale with only $N_j = 128$ data points, GCV does a good job, although the corresponding GCV-function for non-decimated coefficients in Figure 5.7 is clearly smoother. This function is based on $N_j = 2048$ coefficients.

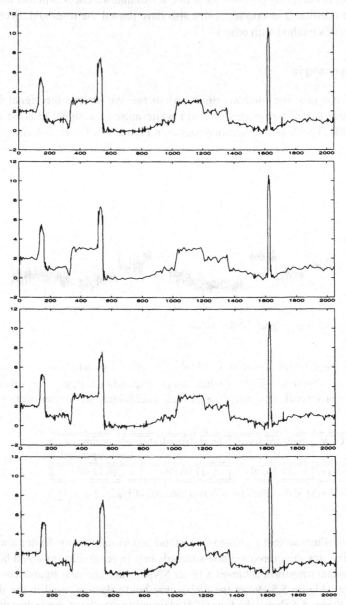

Figure 5.4. Outputs for different schemes, based on GCV threshold estimation. On top: simple, global thresholding the finest four scales. Second plot, tree-structured thresholding as explained in Section 5.3.1 yields a smoother result. The third plot is the output from a level-dependent threshold selection. The fourth one adds to this the use of the non-decimated wavelet transform. All outputs come from an orthogonal Daubechies wavelet transform with three vanishing moments.

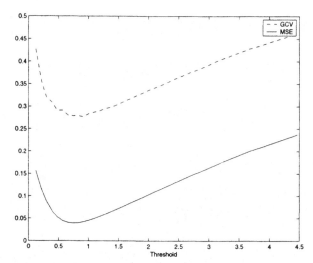

Figure 5.5. GCV and mean square error of the result in function of the threshold λ. This is the selection of one, global threshold for four resolution levels. This threshold is used to produce the first two outputs in Figure 5.4.

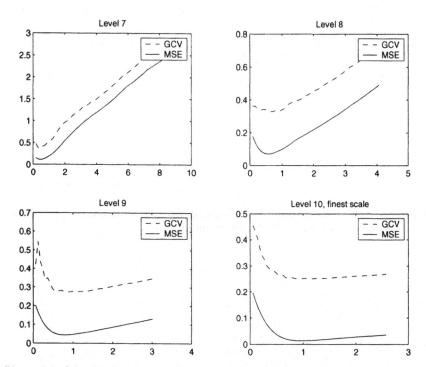

Figure 5.6. GCV_j and mean square error for level-dependent thresholds and a fast wavelet transform. This leads to the third output in Figure 5.4.

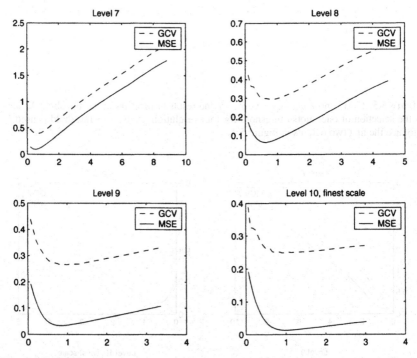

Figure 5.7. GCV_j and mean square error for level-dependent thresholds and a non-decimated wavelet transform. This leads to the bottom plot in Figure 5.4.

5.3.3 Interscale correlations

Classical thresholding takes the coefficient magnitude as a measure of regularity. The soft-threshold operation can be written as

$$
w_{\lambda i} = \begin{cases} 0 & \text{if } |w_i| \leq \lambda \\ \left(1 - \frac{\lambda}{|w_i|}\right) w_i & \text{if } |w_i| > \lambda. \end{cases}
$$

Instead of the local criterion which is the absolute value, the algorithm could consider an alternative measure of regularity m_i. Soft-thresholding on m_i is then:

$$
w_{\lambda i} = \begin{cases} 0 & \text{if } m_i \leq \lambda \\ \left(1 - \frac{\lambda}{m_i}\right) w_i & \text{if } m_i > \lambda. \end{cases}
$$

The same GCV formula as in (4.15) can be used to find a good threshold λ. If $w_i = w_{j,k}$ is the wavelet coefficient at scale j and location k, then an interesting choice for $m_i = m_{j,k}$ is:

$$
m_{j,k} = \prod_{l=j}^{j+D} w_{l,k},
$$

i.e., we consider coefficients at next, coarser scales and compute the correlation between coefficients at the same location. For ease of comparison, it is better to use a non-decimated wavelet transform here: in a non-decimated wavelet representation, it is always clear which coefficient at a next, coarser resolution corresponds to a given coefficient.

The method is similar to an algorithm by Xu, Weaver, Healy and Lu [180], although their procedure does not use a clear notion of threshold. Figure 5.8 compares both methods. We use the same input and wavelet basis as before. Thresholding on correlations is not only superior to the approach by Xu, Weaver, Healy and Lu, it also outperforms all previous methods. Part of the gain is due to the use of non-decimated transforms, but even then there is still an increase in SNR from 18.12 dB up to 19.59 dB. The correlation depth was taken $D = 1$. The method by Xu e.a. achieves a SNR of 18.44 dB when using a correlation depth equal to $D = 2$.

5.4 Intrascale dependencies

Except for Haar wavelets, neighboring basis functions within one scale have non-disjoint support. As a consequence, one singularity in the input signal affects the magnitudes of several consecutive coefficients at each scale. *Block thresholding* methods [91, 34, 35] incorporate information on neighboring coefficients when deciding whether a coefficient is to be kept or to be removed. Moreover, this decision is taken not for a single coefficient, but rather for a group of adjacent coefficients,

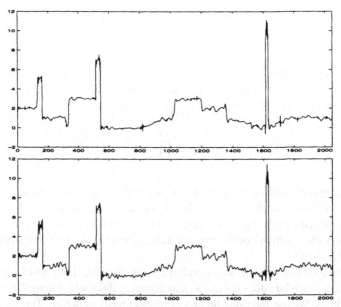

Figure 5.8. Outputs for different schemes, based on interscale correlations as measure of co-efficient importance. The figure on top computes correlations between two successive scales $(D = 1)$ and removes coefficients with a correlation below a threshold. This threshold is selected by a GCV method. Coefficients with a correlation above the threshold are shrunk, in a way similar to soft-thresholding. SNR is 19.59 dB. The figure on the bottom is the output of an algorithm by Xu e.a.: SNR is 18.44 dB.

called a block. The complete block is either shrunk with one common shrinking factor, either totally preserved or totally removed.

Let $I \subset \{1, \ldots, N\}$ denote the set of indices constituting a block of coefficients which are being processed simultaneously. If $\bar{I} = I \cup \partial I$ includes some additional coefficients on which the decision is based, we have as output $w_{\lambda I}$:

$$w_{\lambda I} = \begin{cases} 0 & \text{if } \overline{w_{\bar{I}}^2} \leq \lambda^2 \\ \left(1 - \frac{\lambda^2}{w_{\bar{I}}^2}\right) w_I & \text{if } \overline{w_{\bar{I}}^2} > \lambda^2. \end{cases}$$

In this expression $\overline{w_{\bar{I}}^2}$ stands for the block average energy:

$$\overline{w_{\bar{I}}^2} = \frac{\|w_{\bar{I}}\|^2}{\#\bar{I}}.$$

Choosing $\lambda = 2.1226\sigma$ and $\#I = \log N$ leads to an estimator which asymptotically comes within a constant factor of the ideal (oracle) risk [34].

If $\#I = 1$, blocks reduce to single coefficients, but the decision about the shrinking factor still depends on neighboring coefficients. This can be combined with the universal threshold, but it is interesting to verify what comes out if we apply this block procedure to minimum MSE (or GCV) thresholds. Figure 5.9 has the output for the noisy signal in Figure 5.3. We are again processing four scales of a decimated

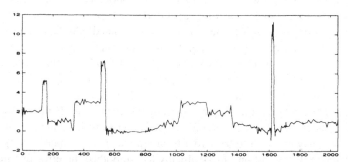

Figure 5.9. Output for block thresholding algorithm, based on level dependent estimations of minimum MSE thresholds. Instead of shrinking each coefficient separately, the shrinking factor is based on the average squared of three successive coefficients. SNR is 17.62 dB. This plot is to be compared with the third plot of Figure 5.4.

transform using the same orthogonal wavelets with three vanishing moments from Daubechies' family. At each scale we estimate the minimum MSE threshold, using the GCV procedure, as for the third plot in Figure 5.4. Instead of simple soft thresholding, we now threshold on the squared average of three consecutive coefficients. This indeed eliminates a part of spurious 'blips', but the result is not so spectacular as for the interscale correlations. This is of course partly due to the fact we are using a decimated transform here, whereas the very nature of the algorithm for interscale correlation thresholding led to a non-decimated approach. On the other hand, the block threshold approach attains a signal-to-noise ratio of 17.62 dB, to be compared with 17.06 dB for level dependent coefficient thresholding. The interscale correlation approach adds 1.47 dB to the 18.12 dB of the simple non-decimated threshold procedure.

Interscale correlations are more important than intrascale dependencies, at least for one dimensional signals. This picture changes dramatically in two dimensions, where edges play an important role. Chapter 6 explores this issue more deeply and looks for appropriate solutions.

5.5 Hard- and soft-thresholding revisited

5.5.1 Risk for hard and soft-thresholding

Section 2.7.4 explained that shrinking coefficients above the threshold reduces the effects of large, spurious noise coefficients. Indeed, if we apply hard-thresholding, using the same threshold as for Figure 5.4 on top, we get the plot in Figure 5.10 on top. Nevertheless, this is far from the optimal hard threshold. Since we know the uncorrupted data, we can easily compute this threshold. While the optimal soft threshold in this example is $\lambda_S^* = 0.75$, the optimal hard threshold is substantially larger: $\lambda_H^* = 1.47$. The second plot in Figure 5.10 shows that the result is quite smooth. Evidently soft-thresholding with the optimal hard threshold leads to an even smoother result, but here bias comes in, as illustrated in the third plot of Figure 5.10. As discussed in Section 3.1.2, smoothness and bias are related: although a non-smooth solution may be biased, it generally holds that too much smoothing causes bias. Figure 5.11 compares the exact risk functions for soft- and hard-thresholding. Only for small thresholds, soft-thresholding is preferable. When using large thresholds, the effect of shrinking coefficients above the threshold becomes important. The optimal hard-threshold performs slightly better than the optimal soft-threshold. On the other hand, the plot makes clear that hard-threshold risk is not a smooth, monotone function. We may expect it to be harder to minimize or to estimate. Taking the expected value in the definition of risk already has smoothed the error function: it is still harder to deal with the MSE function.

5.5.2 Origin and character of the error for hard and soft thresholding

Figure 5.12(a) plots the risk contributions $r_S(v, \lambda_S^*)$ and $r_H(v, \lambda_H^*)$ in one coefficient for soft thresholding and hard thresholding as a function of the uncorrupted value v. It compares the results for the optimal soft threshold and optimal hard threshold. Soft thresholding causes more error in the large coefficients, since these coefficients are unnecessarily shrunk. This is bias, as illustrated from Figure 5.12(b). On the other hand, hard thresholding is more erroneous for coefficients with magnitudes in the neighborhood of the threshold value. These coefficients are on the

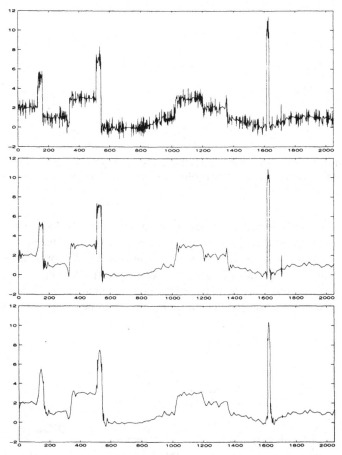

Figure 5.10. Outputs for hard-thresholding, using the optimal soft threshold (top) and the optimal hard threshold (middle), and output for soft-thresholding, using the optimal hard threshold (bottom). Also compare with Figure 5.4. Although soft-thresholding always leads to a smoother result, the optimal hard thresholding procedure yields a fair output. Unfortunately the minimum MSE for hard-thresholding is hard to find.

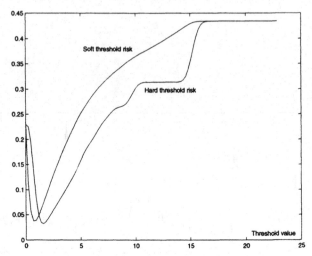

Figure 5.11. Exact risk function for hard and soft thresholding, applied to the wavelet coefficients of the signal in Figure 3.4, using Daubechies' orthogonal wavelets with three vanishing moments.

boundary between large and small, and the error here is mostly due to variance: little variation in the input may cause an important uncertainty in the hard threshold output: a coefficient may be thrown away or kept entirely. Moreover, since the optimal hard threshold is twice as large as the optimal soft threshold, coefficients in this region of magnitude are more biased by hard thresholding than by soft thresholding. These intermediate coefficients do not appear in intervals where the signal

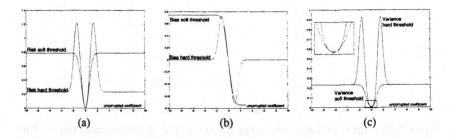

Figure 5.12. Risk, bias and variance for soft thresholding and hard thresholding, as a function of the uncorrupted coefficient value. Hard thresholding better preserves large coefficients, but introduces more variance at intermediate coefficient values.

is smooth, nor do they correspond to the very location of a singularity. These coefficients appear in the neighborhood of singularities: this explains the oscillatory behavior of the hard threshold output in the neighborhood of signal singularities in Figure 5.10(middle). So, the origin and the nature of these Gibbs-like phenomena is different from that of the spurious blips which appear in a soft threshold reconstruction. As a consequence, techniques to reduce blips (such as tree structured or correlation based thresholding) do not apply here.

If the wavelet transform is orthogonal and the noise is stationary and white, we can compute the covariance matrix S_λ after thresholding in the wavelet domain: it is a diagonal matrix, with diagonal elements:

$$\sigma_i^2 = r(v_i, \lambda) - E\omega_\lambda^2.$$

The variances of the output samples are then the diagonal elements of the matrix

$$Q_\lambda = WS_\lambda W^T,$$

with W the inverse wavelet transform. We can compute these variances (with an algorithm that can be optimized up to linear complexity) together with the bias after reconstruction:

$$E\eta_\lambda = WE\omega_\lambda.$$

A detailed plot of the expected output Ey_λ with its $[-3\sigma, 3\sigma]$ confidence interval appears in Figure 5.13. We should be careful in interpreting these 'confidence' intervals, since there is no guarantee at all that the noise after reconstruction is normally distributed. As a matter of fact, it is not even symmetrically distributed. Nevertheless, plotting the standard deviation gives an indication for the uncertainty in each reconstructed point. This confirms our analysis: hard threshold reconstruction has more difficulties to deal with singularities.

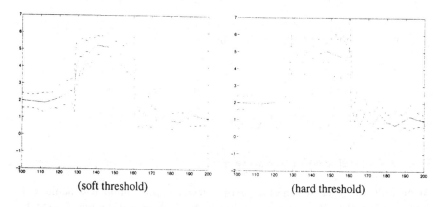

(soft threshold)	(hard threshold)

Figure 5.13. Expected reconstruction and uncertainty intervals for soft and hard thresholding. Hard threshold reconstruction has more difficulties to deal with singularities, giving rise to Gibbs-like phenomena.

5.5.3 Estimation of the optimal hard threshold

The previous analysis motivates the use of intermediate shrinking rules, such as described in several publications [31, 84, 13]: these approaches are continuous and thereby keep the variance of the output under control, and on the other hand, large coefficients are preserved, thereby reducing bias. On the other hand, the difference in quality between soft and hard thresholding is rather small. Looking for alternative selection criteria may lead to more important improvements.

In many cases, the 'keep-or-kill' principle of hard thresholding is easy to extend. Therefore, we are interested in the optimal hard threshold. Since the GCV procedure is not appropriate for estimating the hard threshold risk function, we proceed in two steps. We first compute the optimal soft threshold. From this we could find a hard threshold on a heuristical basis. For instance, we could look for the hard threshold that removes the same amount of energy as the obtained soft threshold.

We propose a less heuristical approach, starting from the exact risk calculus. It is easy to check that the risk contribution of one coefficient satisfies:

$$r_H(v, \lambda) - r_S(v, \lambda) = 2\sigma^2\lambda \left(\phi(\lambda + v) + \phi(\lambda - v)\right) - 2\lambda^2 \mathrm{P}(|v + \omega| > \lambda).$$

An unbiased estimate for the second term is trivial: we count the coefficients above the threshold. From experiments we conclude that a fair and not too biased estimate for the first term in this expression comes from plugging in the soft-thresholded coefficients at the place of v.

So, if we know the soft-threshold error $R_S(\lambda)$ or its estimate (e.g. by cross validation), we estimate the hard-threshold error as:

$$\hat{R}_H(\lambda) = R_S(\lambda) + 2\hat{\sigma}^2\lambda\frac{1}{N}\sum_{i=1}^{N}\left(\phi(\lambda + w_{\lambda i}) + \phi(\lambda - w_{\lambda i})\right) - 2\lambda^2\frac{N - N_0}{N}.$$

A plot of this estimate appears in Figure 5.14.

5.6 Test examples and comparison of different methods

We now discuss a couple of test examples.

5.6.1 Orthogonal transform, white noise

We have already discussed an example of a piecewise polynomial with additive, stationary, white, and Gaussian noise in Section 5.3.2. A second example is Donoho's and Johnstone's 'HeaviSine' signal [78]:

$$f(x) = 4\sin(4\pi x) - \mathrm{sign}(x - 0.3) - \mathrm{sign}(0.72 - x),$$

sampled at $N = 1024$ equidistant points and corrupted by white, stationary noise with $\sigma = 0.5$. Again we use Daubechies wavelets with three vanishing moments, and this time we process six scales. This leads to the same test conditions as in [20]. Figure 5.15 has the noise-free data and the noisy input. The following figure and Table 5.2 summarize the results. Comparison of the different algorithms leads to

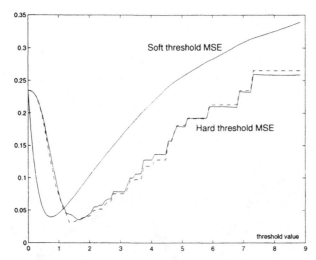

Figure 5.14. Estimation of hard threshold MSE function. The estimate is in full line, the true hard threshold MSE is in dashed line. Once more, we note that the hard threshold MSE function is far from smooth, as compared to the soft threshold MSE.

similar results. We note that the underlying signal is smoother than in the previous example. Using a smooth wavelet basis, like the Daubechies basis with three vanishing moments, performs better in such cases than in the rather blocky signal of the previous example.

simple GCV	tree-structured	level-dependent	non-decimated, level-dependent
25.60 dB	27.17 dB	25.31 dB	27.95 dB

Table 5.2. Output SNR-values for different methods of Figure 5.16.

The two examples illustrate that level-dependent thresholding for decimated wavelet coefficients is not always that useful: the signal characteristics do not depend too much on scale in these examples. Conclusions may be different for other signals, and certainly for data with correlated noise. In that case, level-dependent thresholding is absolutely necessary and it acts as a whitening filter. This is illustrated in two examples with images.

5.6.2 Biorthogonal transform, colored noise

To the image of Figure 3.7 we add artificial colored noise. This noise was the result of a convolution of white noise with a FIR-highpass-filter. The signal-to-noise ratio is 4.97 dB. As wavelet filter, we use the variation on the CDF-(spline)-filters "with less dissimilar lengths" [49, 18]. We choose a basis with four primal and four dual

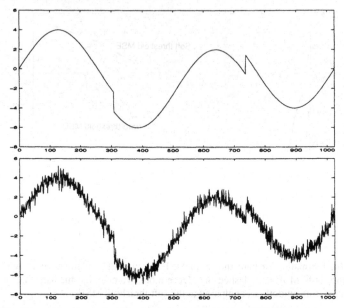

Figure 5.15. 'HeaviSine' test signal and noisy version SNR = 15.47 dB.

vanishing moments. Figure 5.17 shows that the algorithm achieves a signal-to-noise ratio of 16.83 dB. Figure 5.18 plots the GCV-function and the mean square error for the vertical component at the one but finest resolution level. Table 5.3 compares the thresholds for different procedures. The first column contains the results for level-dependent GCV. This has to be compared with the thresholds minimizing the mean square error (MSE) in terms of wavelet coefficients. For an orthogonal transform, the corresponding SNR-value would have been the absolute maximum. Since we work with non-orthogonal transforms, and maximize in the wavelet domain, the value of 17.00 dB is only an approximative maximum. We add the results for SURE and universal thresholding. These algorithms need an explicit variance estimator. We use $\hat{\sigma}_j = \text{MAD}(w_{j,k}, k = 1 \ldots 2^j)/0.6745$, where MAD is the median absolute deviation. If we suppose full knowledge of the noise energy in each component and at each level, the SURE-procedure rises from 16.24 dB to 16.94 dB, which is slightly better than the GCV based method. Both the GCV and the SURE procedure remove all coefficients at the finest scale ($j = 7$): the thresholds are equal to the largest coefficient. We remark that the universal threshold can be seen as a "statistical upper bound": if $N \to \infty$, it is almost sure that a pure noise coefficient is removed. As discussed in Section 3.4.6, this over-smoothing threshold is not appropriate for image processing. We also note that we only threshold coefficients at the three finest resolution levels. Coarse levels contain more important image information and thresholding these coefficients may cause a considerable bias and introduce visual artifacts. Figure 5.19 illustrates the smoothing effect of the

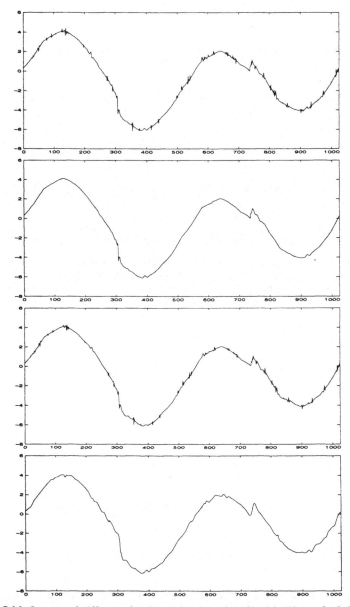

Figure 5.16. Outputs of different algorithms for the noisy signal in Figure 5.15. From top to bottom: (1) global thresholding the finest six scales. (2) tree-structured thresholding. (3) output from a level-dependent approach. (4) level-dependent thresholds for a non-decimated wavelet transform. All outputs come from an orthogonal Daubechies wavelet transform with three vanishing moments.

Figure 5.17. Left: an image with artificial, correlated noise. The noise is the result of a convolution of white noise with FIR highpass filter. Right: the result after level-dependent wavelet thresholding. We use biorthogonal filters with four primal and four dual vanishing moments and filter lengths 7 and 9. Signal-to-noise ratio rises from 4.97 dB to 16.83 dB.

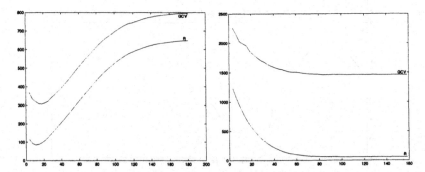

Figure 5.18. Mean square error and Generalized Cross Validation for vertical component coefficients at the one but finest resolution level (Left) and at the finest resolution level (Right).

	GCV	MSE	SURE	universal
$\lambda_{7,\text{hor}}$	143.7	92.51	143.7	171.9
$\lambda_{7,\text{ver}}$	148.6	118.7	148.6	168.0
$\lambda_{7,\text{diag}}$	125.4	156.7	125.4	171.5
$\lambda_{6,\text{hor}}$	10.21	10.40	22.89	72.82
$\lambda_{6,\text{ver}}$	15.98	11.53	19.67	66.93
$\lambda_{6,\text{diag}}$	63.72	63.58	63.72	116.7
$\lambda_{5,\text{hor}}$	0.7780	1.677	18.66	85.82
$\lambda_{5,\text{ver}}$	7.465	2.026	13.78	63.92
$\lambda_{5,\text{diag}}$	13.13	5.282	15.90	52.16
SNR	16.83	17.00	16.24	12.64

Table 5.3. Comparison of thresholds for different procedures. The first column contains the results for level-dependent GCV. This is to compare with the thresholds minimizing the mean square error (MSE) in terms of wavelet coefficients. We add the results for SURE and universal thresholding.

redundant transform: the reconstruction from the overcomplete data representation reduces visual artifacts. Signal-to-noise ratio is now 18.02 dB.

Figure 5.19. Result of level dependent wavelet thresholding on the redundant wavelet transform of the image with noise in Figure 5.17. We use the same wavelet filters. Signal-to-noise ratio is now 18.02 dB.

We now apply the method to a realistic image. Figure 5.20 represents an aerial photograph of 340×350 pixels. We use the same biorthogonal filters with four primal and four dual vanishing moments as in the previous example. Table 5.4 compares the different thresholds of GCV with SURE. In contrast to the previous, artificial example, the threshold values are quite different at coarse levels. The SURE-thresholds are too high, probably because the MAD-estimator fails for this example.

	GCV	SURE		GCV	SURE
$\lambda_{8,hor}$	11.06	11.83	$\lambda_{6,hor}$	10.79	26.37
$\lambda_{8,ver}$	14.13	14.93	$\lambda_{6,ver}$	12.07	19.56
$\lambda_{8,diag}$	10.44	10.44	$\lambda_{6,diag}$	12.32	19.90
$\lambda_{7,hor}$	9.22	14.65	$\lambda_{5,hor}$	12.34	75.59
$\lambda_{7,ver}$	12.41	15.43	$\lambda_{5,ver}$	12.36	37.32
$\lambda_{7,diag}$	16.94	16.94	$\lambda_{5,diag}$	13.81	41.84

Table 5.4. Comparison of different threshold values for GCV and SURE, applied on coefficients at four scales of the image in Figure 5.20. We call $j = 8$ the finest scale. Both methods show quite different thresholds at coarse scales. The SURE-thresholds are too high, probably because the variance estimator fails. This illustrates the advantage of an automatic threshold estimator.

Figure 5.21 contains the result for the GCV-procedure. The four finest resolution levels are thresholded. As can be expected, the algorithm does not distinguish real noise from the apparently noisy texture in the foliage of the trees.

Figure 5.20. Aerial photograph with noise (340 × 350 pixels).

Figure 5.21. Result of level-dependent wavelet thresholding for the aerial photograph.

5.7 Integer wavelet transforms

In applications like digital image processing, the input data are often integers. Section 2.6.3 explains that an integer wavelet transform avoids floating point operations and storage. If the input is affected by noise, but still integer, this noise cannot take an arbitrary real value and so its distribution cannot be Gaussian. Moreover, the integer wavelet transform is non-linear and so it does not preserve additivity nor stationarity. All these conditions are stricto sensu necessary for a correct use of a GCV-threshold estimation.

Nevertheless, an artificial test example illustrates that, in practice, these conditions do not pose serious problems. Figure 5.22(a) shows a noise-free DSA (Digital Subtraction Angiography) test image. In Figure 5.22(b) we add artificial, colored noise. This noise was the result of a convolution of white noise with a FIR high pass-filter. The signal-to-noise ratio is 10 dB.

We compute the redundant, integer wavelet transform of the noisy and the noise-free image and estimate the optimal threshold at each resolution level and for each component by the GCV-procedure. Since we know the noise-free wavelet coefficients, we can compare the GCV-function with the mean square error as a function of the threshold. Figure 5.23(a) shows this comparison for the vertical component at the finest resolution level. Both $GCV_j^c(\lambda)$ and $R_j^c(\lambda)$ have about the same minimum. Figure 5.23(b) compares both functions at the one but finest resolution level. The optimal threshold at this level is close to zero. This is not surprising: we have

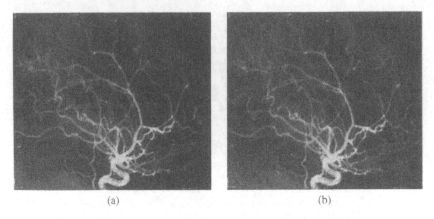

(a)   (b)

Figure 5.22. An artificial test example. (a) a noise-free DSA test image. (b) The same image with artificial, additive and correlated noise. The noise is the result of a convolution of white noise with a FIR high pass filter. Signal-to-noise ratio is 10 dB.

added high-frequency noise which mainly manifests at fine scales. As Figure 5.24b shows, thresholding at this level causes considerable blur and artifacts and loss of important details: small blood-vessels become very unclear or even completely disappear. We get a better result if we only apply the algorithm at the finest resolution level. Figure 5.24a shows this reconstruction: signal-to-noise ratio is now 19.94 dB, compared to 17.27 dB for a threshold at two levels.

Table 5.5 compares the integer GCV procedure with other, classical threshold methods. It shows that, at least for this example, using the integer transform instead of the classical, linear transform poses no problem. (In this case, there is even a slight improvement.) The table also illustrates that the GCV-method performs at least as well as other threshold selection procedures, although GCV does not use information on the amount of noise (the deviation σ_j^c). The SURE- and universal procedures do need the values of σ_j^c. For the values in this table, we used the exact σ_j^c.

integer GCV	classical GCV	SURE	universal
19.94	19.88	19.26	18.72

Table 5.5. Comparison of different threshold procedures, applied to the finest scale of the co-efficients of Figure 5.22. In all cases, we used a redundant transform with Cohen-Daubechies-Feauveau (2,2)-filters.

Our last example is an MRI image with 'natural' (no artificial) noise. This image has 128 by 128 pixels and shows a human knee. The input is in Figure 5.25(a). Figure 5.25(b) contains the result of the de-noising algorithm for a fast wavelet transform. Figure 5.25(c) is the result for a non-decimated transform. Figures 5.25(d)

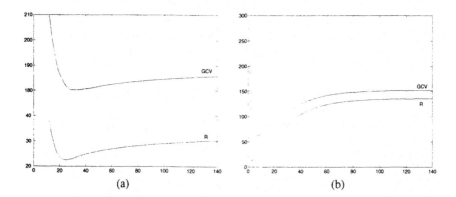

Figure 5.23. An artificial test example: a comparison of $GCV_j^c(\lambda)$ and $R_j^c(\lambda)$ for the vertical component at the (a) finest resolution level and (b) one but finest resolution level. Both $GCV_j^c(\lambda)$ and $R_j^c(\lambda)$ have about the same minimum. At the one but finest level, the optimal threshold is close to zero, which indicates that the noise at this level is neglectable.

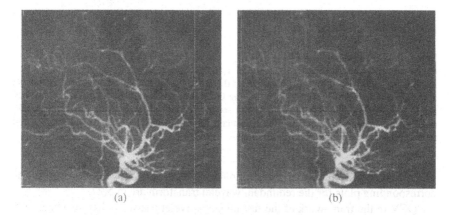

Figure 5.24. An artificial test example: reconstruction by inverse redundant transform after removing small coefficients at (a) the finest resolution level only (Signal-to-noise ratio is 19.94 dB.), and (b) the two finest resolution levels. Signal-to-noise ratio is 17.27 dB. Thresholding at coarse levels introduces more visible artifacts.

and 5.25(e) show the results if one uses universal thresholds. This illustrates, once more, that the universal threshold is in fact not appropriate for image de-noising.

We used biorthogonal CDF(2,2)-wavelet filters [49]. This is one of the popular filters for image processing. Its decomposition into lifting steps is particularly easy [164]. In principle, the success of GCV in estimating the optimal threshold does not depend on the choice of a wavelet basis. Figure 5.26 shows the GCV-functions

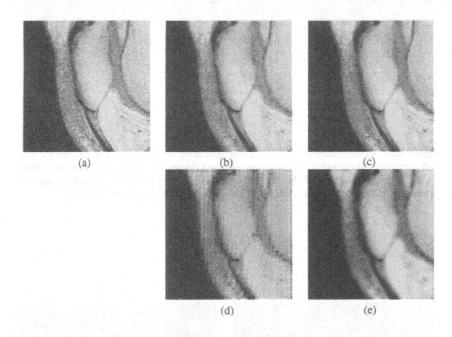

Figure 5.25. An example. (a) The input image, an MRI image (128 × 128 pixels) with noise. (b) Result after thresholding the fast wavelet coefficients at the first and second resolution level, using GCV-thresholds. (c) Result after thresholding the redundant wavelet coefficients at the first and second resolution level, using GCV-thresholds. (d) Result after thresholding the fast wavelet coefficients at the first and second resolution level, using universal thresholds. (e) Result after thresholding the redundant wavelet coefficients at the first and second resolution level, using universal thresholds.

of the first (finest) and second resolution level of the fast wavelet transform. The corresponding plots for the redundant wavelet transform are in 5.27.

GCV in the framework of the fast integer wavelet transform has also been successfully applied for noise reduction in large-scale images [166, 168].

5.8 Non-Gaussian noise

So far, we have limited discussion to additive, stationary, Gaussian noise. Evidently, the method still works fine for Gaussian-like distributions. Most distributions are

LH   HL   HH

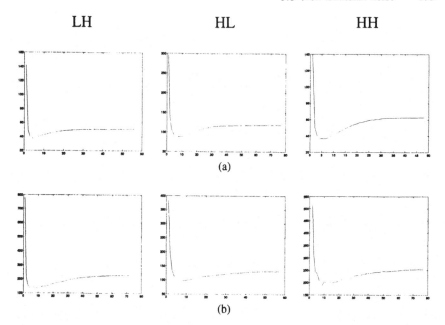

(a)

(b)

Figure 5.26. GCV-functions for a fast wavelet transform of MRI-image of figure 5.25. (a) The three components at the finest resolution level. (b) The three components at the second resolution level.

not invariant under linear transforms, such as the wavelet transform, so it would be hard to calculate the wavelet coefficient density function exactly. If the input noise has finite variance (and zero mean), the normal density is generally a good approximation. Other types, like shot-noise (also known as salt-and-pepper noise), were not treated. Shot noise does not cause the typical small, noisy coefficients.

An important class of heteroscedastic noise is *multiplicative* or *Poisson* noise:

$$P(y_i|f_i) = \frac{e^{-f_i} f_i^{y_i}}{y_i!}.$$

The noisy data can only take integer values. This is a good model in situations where intensity (image grey values) are proportional to the result of counting incoming light particles. CT (computer tomography) scanning is an example of this situation. This model also appears in some algorithms for statistical density estimation. Suppose we want to estimate the density $f_X(x)$ of some random variable X, and we have n observations. If we denote by y_i the number of evaluations that fall between x_{i-1} and x_i, $i = 1, \ldots, N$, then

$$f_i := Ey_i = n \int_{x_{i-1}}^{x_i} f_X(x)dx,$$

and y_i has an approximate Poisson distribution (The exact distribution is a heteroscedastic binomial).

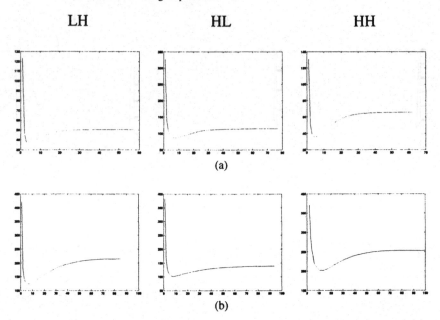

Figure 5.27. GCV-functions for a redundant wavelet transform of MRI-image of figure 5.25. (a) The three components at the finest resolution level. (b) The three components at the second resolution level.

Most density estimation algorithms proceed in a different way: they start from unbiased estimates for scaling coefficients f_k at the finest scale. The density function is expanded as:

$$f_X(x) = \sum_k f_k \varphi_{J,k}(x),$$

with:

$$f_k = \int_{-\infty}^{\infty} \tilde{\varphi}_{J,k}(x) f_X(x) dx,$$

and the initial, noisy estimates are:

$$y_k = \sum_{l=1}^{n} \tilde{\varphi}_{J,k}(X_l).$$

In a Haar basis, this corresponds to simply counting the number of observations in subsequent intervals, but even in the general case, the assumption of normality does not hold. The heteroscedastic character of Poisson(-like) noise makes it difficult to remove it by one threshold, even if this threshold is scale-adaptive. Anscombe's [10] transformation

$$z_i = 2\sqrt{y_i + 3/8}$$

yields data with a distribution closer to the Gaussian. Using this transformation as a preprocessing step allows for a more successful application of thresholding. Figure 5.28 gives an example.

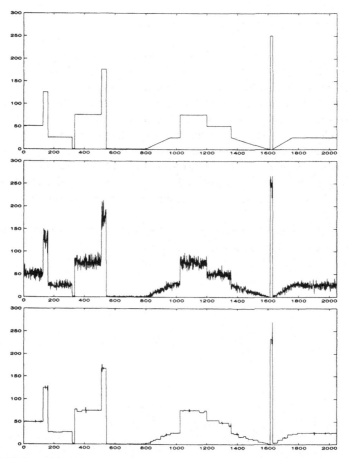

Figure 5.28. Smoothing Poisson data, using Anscombe's pre-processing. Figure on top: plot of the expected number of hits. This is the uncorrupted data. In the middle: plot of the occured number of hits. These data are Poisson distributed (SNR = 15.93 dB). The bottom plot contains the smoothed data, using Anscombe's pre-processing followed by wavelet thresholding. The algorithm uses one global threshold, estimated by GCV, applied to the Haar transform of the input. (SNR = 22.5 dB)

Intuitively, one can understand this as follows: if y_i is Poisson distributed, we know that $Ey_i = f_i = \sigma_{y_i}^2$. As a consequence, the vector $y_i/\sqrt{f_i}$ has a constant variance, equal to one. This can be approximated by $y_i/\sqrt{y_i} = \sqrt{y_i}$. This analysis makes clear that peaks of important intensity have the largest variances as well and therefore are subject to the largest rescaling factors. As a consequence, Anscombe's pre-processing shows a tendency to blur sharp, brief structures [112].

Alternatively, we could use expression (2.20) to estimate the covariance matrix in the wavelet domain:

$$\hat{S} = \tilde{W}\hat{Q}\tilde{W}^T,$$

where $\hat{Q}$ is a diagonal matrix, with diagonal elements:

$$\hat{Q}_{ii} = \hat{\sigma}^2_{y_i} = y_i.$$

$\hat{S}$ is an unbiased estimate for the covariance matrix of the wavelet coefficients. If the input noise is uncorrelated, computation of $\hat{S}$ has linear complexity. Rescaling the coefficients of the non-pre-processed data as in (5.1) leads to homoscedastic noise in the wavelet domain. Figure 5.29 illustrates that the coefficients corresponding to signal features are still clearly visible. The noise is of course non-Gaussian, even after rescaling. For instance, the algorithm did not correct for skewness of the noise densities. Figure 5.30 however shows that this has only a moderate effect on the GCV procedure. Evidently, other threshold selection procedures are are possible too [112].

Automatically selected thresholds for other types of heteroscedastic and locally stationary noise [173] merit further investigation.

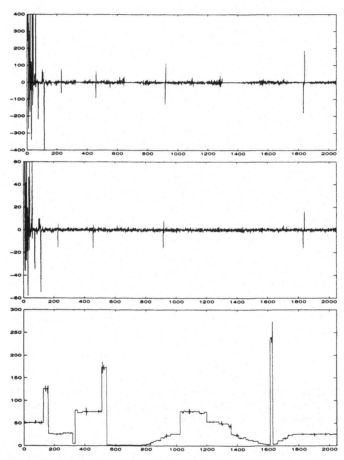

Figure 5.29. Smoothing Poisson data, using variance corrected thresholds. Figure on top: plot of the wavelet coefficients of the noisy signal in Figure 5.28. The noise is clearly heteroscedastic. The plot in the middle shows the wavelet coefficients after rescaling with an unbiased estimate of their standard deviation. This is homoscedastic noise. Shrinking these coefficients with a GCV threshold leads to the curve in the bottom plot. SNR = 23.8 dB.

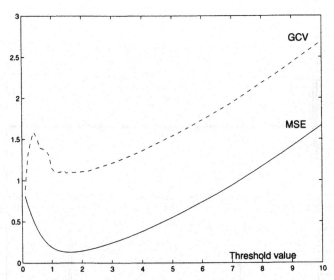

Figure 5.30. *GCV* and mean square error for wavelet coefficients of the middle plot in Figure 5.29. These coefficients have homoscedastic but non-Gaussian noise. This non-Gaussian distribution has however no substantial influence.

6

Bayesian correction with geometrical priors for image noise reduction

6.1 An approximation theoretic point of view

Image processing is not merely a two-dimensional translation of traditional signal processing techniques. The two-dimensional character has some important consequences, such as the existence of line singularities, manifesting as edges. The observations explained in this section also provide the basis for the development of new types of basis functions, such as ridgelets [37].

6.1.1 Step function approximation in one dimension

Suppose we want to approximate the step function of Figure 6.1. A periodic extension of this function can be decomposed into a Fourier series:

$$f(x) = \sum_{k \in \mathbb{Z}} a_k e^{-i2\pi kx}.$$

The equation sign indicates convergence, both pointwise as in $L_2[0,1]$-norm. The coefficients a_k depend of course on the precise position of the singularity, but they behave like:

$$a_k = \mathcal{O}\left(\frac{1}{|k|}\right).$$

We use this Fourier expansion to *approximate* the step function by taking the $2n+1$ harmonics with index $k = -n, \ldots, n$:

$$f_{2n+1}(x) = \sum_{k=-n}^{n} a_k e^{-i2\pi kx}.$$

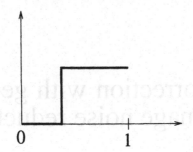

Figure 6.1. Step function.

Since the coefficients decrease for $|k| \to \infty$, this *linear* approach happens to coincide with taking the largest contributions. This Fourier basis is orthonormal, so the approximation error satisfies

$$\|\epsilon_{2n+1}\|^2_{L_2[0,1]} \;=\; \|f - f_{2n+1}\|^2_{L_2[0,1]} = \sum_{|k| \geq n+1} |a_k|^2$$

$$=\; \mathcal{O}\left(2 \sum_{k=n+1}^{\infty} \frac{1}{k^2} \right) = \mathcal{O}(n^{-1}).$$

We may conclude that a one dimensional Fourier decomposition performs as:

$$\|\epsilon_n\| = \mathcal{O}(n^{-1/2}).$$

This bad approximation of piecewise smooth signals is a well known drawback of the Fourier decomposition. The reason is of course that all basis functions cover the entire interval, and so all of them get in touch with the singularity, all of them have a contribution to it.

This is not the case in a wavelet decomposition, where at each scale only a constant number of coefficients are non-zero. For the Haar transform, there is only one function, say ψ_{j,k_j} with a non zero coefficient w_{j,k_j}. By orthogonality, the approximation error of the orthogonal projection f_J of f on V_J equals:

$$\|\epsilon_J\|^2_{L_2[0,1]} = \|f - f_J\|^2_{L_2[0,1]} = \sum_{j=J+1}^{\infty} w^2_{j,k_j}.$$

If $|f(x)| \leq 1$, we have

$$|w_{j,k_j}| \leq \int_0^1 |\psi_{j,k_j}(x)| \, dx = 2^{-j} 2^{j/2}.$$

Hence

$$\|\epsilon_J\|^2_{L_2[0,1]} = \mathcal{O}\left(\sum_{j=J+1}^{\infty} 2^{-j} \right) = \mathcal{O}(2^{-J}).$$

This expresses that a wavelet basis indeed captures isolated singularities much more efficiently than does a Fourier basis. We do not know in advance which coefficients are going to be non-zero: this depends on the input signal and more precisely on the exact position of the jump. Therefore, keeping the non-zero wavelet coefficients is a *non-linear* approximation.

If we have a superposition of this step function f and a C^α smooth function g, none of the wavelet coefficients of $h = f + g$ is exactly zero. The smooth part g is best approximated with a linear approach: take all coefficients up to scale J. If the number of vanishing moments $p \geq \alpha$, the approximation error satisfies [130]:

$$\|\epsilon_{g,J}\| = \|g - g_J\| = \mathcal{O}(2^{-J\alpha}).$$

This approximation uses 2^{J+1} coefficients. If we want the same order of precision for the non-smooth component, we add $\lceil J\alpha \rceil$ coefficients of f in the non-linear way, and the overall approximation error $\epsilon_h(x) = \epsilon_f(x) + \epsilon_g(x)$ is then bounded by:

$$\|\epsilon_h\| \leq \|\epsilon_f\| + \|\epsilon_g\| = \mathcal{O}(2^{-J\alpha}).$$

This approximation uses $2^{J+1} + \lceil J\alpha \rceil = \mathcal{O}(2^J)$ coefficients. If we call this number n, we conclude that the error of a one-dimensional wavelet approximation behaves as

$$\|\epsilon_n\| = \mathcal{O}(n^{-\alpha}),$$

for smooth as well as for piecewise smooth functions. Isolated singularities have no influence on the asymptotic approximation error.

6.1.2 Approximations in two dimensions

Now suppose we are given a two-dimensional function $f(x, y) \in L_2[0, 1]^2$, which is 0 in one part of the square and 1 in the other part. The boundary between these two parts is a simple line, as in Figure 6.2. The coefficients of the Fourier expansion

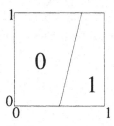

Figure 6.2. Two-dimensional step function.

of this function

$$f(x, y) = \sum_{k \in \mathbb{Z}} \sum_{l \in \mathbb{Z}} a_{k,l}\, e^{i2\pi kx}\, e^{i2\pi ly}$$

can be found as:

$$
\begin{aligned}
a_{k,l} &= \int_0^1 \int_0^1 f(x,y)\, e^{-i2\pi kx} e^{-i2\pi ly}\, dx\, dy \\
&= \mathcal{O}\left(\frac{1}{|kl|}\right).
\end{aligned}
$$

A linear approximation keeps all coefficients with indices k and l such that $|k|+|l| \leq m$, for a given m. Figure 6.3 shows the position of these indices in $\mathbb{Z}^2$ for $m = 3$. The approximation error satisfies:

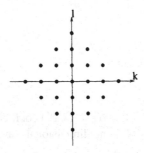

Figure 6.3. Position of indices in $\mathbb{Z}^2$ corresponding to coefficients in a linear Fourier approximation.

$$
\|\epsilon_n\|^2 = \sum_{k \in \mathbb{Z}} \sum_{l:\,|k|+|l| \geq m+1} |a_{k,l}|^2 = \mathcal{O}(m^{-1})
$$

Since $n = (2m + 1)^2$, we have

$$
\|\epsilon_n\| = \mathcal{O}(n^{-1/4}).
$$

This says that for an equal order of magnitude of the error as in the one-dimensional case, we need the square of the number of coefficients.

We now proceed to a (Haar) wavelet expansion. The basis contains three types of functions: horizontally oriented, vertically oriented and diagonally oriented functions as depicted in Figure 6.4. At each scale we have $K2^j$ non-zero coefficients

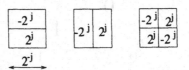

Figure 6.4. Two-dimensional Haar basis functions.

with all three types of basis functions. The exact value of the constant $K \geq 1$ depends on the length of the singularity, i.e. its orientation in the image. If $|f(x,y)| \leq 1$, the coefficients satisfy:

$$|w_{j,k,l}^{\mathrm{HH}}| \leq \int_0^1 \int_0^1 |\psi_{j,k,l}^{\mathrm{HH}}(x,y)|\, dx dy = 2^j\, 2^{-j} \times 2^{-j},$$

and similarly for horizontal and diagonal subbands. The approximation error becomes:

$$\|f - f_J\|^2 \leq 3 \sum_{j=J+1}^{\infty} K 2^j 2^{-j2} = \mathcal{O}(2^{-J}).$$

This is the same order of approximation as in the 1D-case, we now need

$$n = \sum_{j=0}^{J} K 2^j = \mathcal{O}(2^J)$$

coefficients. So the order of approximation is now:

$$\|f - f_n\| = \mathcal{O}(n^{-1/2}),$$

while in the one-dimensional case we had $n = J$ and

$$\|f - f_n\| = \mathcal{O}(2^{-n}).$$

This is not just squaring the number of coefficients to obtain a comparable error in two dimensions. This dramatic change comes from the difference between a point singularity in one dimensional signals and a line singularity in two dimensions. Of course, point singularities also exist in images, but they are far less important, and, after all, line singularities certainly do not exist in one dimensional signals. A point has no dimension, at each scale it only interferes with a fixed number of basis functions. A line has however a certain length. Consequently, the number of basis functions meeting this line increases for finer scales.

For a piecewise smooth function of the form $h = g + f$, with $g \in C^\alpha$, we have $\|g - g_J\| = \mathcal{O}(2^{-J\alpha})$, provided that the wavelet basis has $p \geq \alpha$ vanishing moments. This linear approximation uses $n = \mathcal{O}(2^{2J})$ coefficients, so the order of approximation is $\mathcal{O}(n^{-\alpha/2})$. We need coefficients at $\lceil 2J\alpha \rceil$ resolution levels to represent f with the same the same order of accuracy as g. This means $\mathcal{O}(2^{2J\alpha})$ additional non-zero coefficients. The total number of coefficients to achieve $\|h - h_n\| = \mathcal{O}(2^{-J\alpha/2})$ is then $n = \mathcal{O}(2^{2J\alpha} + 2^J) = \mathcal{O}(2^{2J\alpha})$. A wavelet approximation for a piecewise smooth function in two dimensions thus has an accuracy of $\mathcal{O}(n^{-1/2})$. Unlike in the one-dimensional case, the line singularity does have an important impact on the quality of the wavelet approximation: all the benefits from using more than one vanishing moment seem to be lost.

6.1.3 Smoothness spaces

As discussed in Section 3.7.2, Besov spaces are well characterized by wavelet coefficients: wavelet bases are unconditional bases for these spaces.

On the other hand, members of Besov spaces also show good wavelet approximation properties: if $\|f\|_{B^\alpha_{p,q}} < \infty$, then f can be approximated with an accuracy of $\mathcal{O}(n^{-\alpha/2})$ by n coefficients [68].

Since for a simple image as in the previous section, this order of approximation is $\mathcal{O}(n^{-1/2})$, this seems to suggest that typical images are in Besov spaces with relatively low α, even if the regions between the edges are very smooth. Typical images are reported to live in Besov spaces with values of α between 0.3 and 0.6 [68].

6.1.4 Other basis functions

From the previous analysis, we conclude that wavelets may not provide the ultimate representation for images, and, consequently, Besov spaces may not be the optimal way to describe images. It is of course true that wavelets and Besov spaces *are* successful, but looking for better generalizations of the wavelet idea for more-dimensional applications is an interesting — though difficult — challenge. The previous argument comes from approximation theory, but it has consequences in statistical estimation: a basis that performs well in approximation of piecewise smooth functions is appropriate for noise reduction. This is a motivation for the development of new types of basis functions, like ridgelets [74, 37, 38], or curvelets [39]. Ridgelets, for instance, are well suited for images with straight lines, like the example of Figure 5.17. Other, for instance medical, images however, typically have non-straight edges, or straight edges in some regions and non-straight edges elsewhere. Other singularities may even show different behavior at different scales. For all these reasons, it is far from trivial to find a basis well adapted to all circumstances.

This text takes a different approach: it applies the classical two-dimensional wavelets and concentrates on the coefficients in the basis for a description of edges. This description is based on a random prior model for these coefficients and leads to a Bayesian algorithm, the philosophy of which is described in Section 2.8. The idea is to capture a global structure, such as an edge, by describing local interdependencies of wavelet coefficients. This is interesting from the computational point of view and this approach adapts itself to the singularity at hand.

6.2 The Bayesian approach

6.2.1 Motivation and objectives

In one dimension, as in two dimensions, wavelet basis functions are localized in space and scale (frequency). As a consequence, manipulating a coefficient has a local effect, both in space and frequency. This is an important advantage of wavelet based methods.

On the other hand, usual *classification* rules are local too, and do not take into account all the correlations that exist among neighboring coefficients. Although a wavelet transform has decorrelating properties, this decorrelation is not complete (a wavelet transform is sometimes seen as an *approximation* of a Karuhnen-Loève-transform). We distinguish two types of correlations:

1. Important image features correspond to large coefficients at different scales: these coefficients are of course correlated. This type of correlation is inherent to all wavelet decompositions: it reflects the multiscale nature of it. In Chapter 5 we proposed, cited and discussed a couple of *deterministic* algorithms that take this multiresolution character into account. Other algorithms start from different variations of a stochastic 'tree' model for uncorrupted wavelet coefficients in a multiscale structure [44, 57, 122].

2. The second type of correlation is within one scale and is specific for two-dimensional inputs, like images: important coefficients tend to be clustered on the location of edges.

We assume that classical thresholding, possibly extended to deal with interscale correlations, performs sufficiently well for the first type of intercoefficient dependencies. This chapter concentrates on the second type of correlations. For the stochastic description of these structures, we need *geometrical* prior models. This leads to a multiscale version of *Markov Random Field Models* (MRF). Similar approaches are in [22, 27, 125, 124, 123].

The Bayesian approach aims at two objectives at once: by taking into account the geometrical structures in the coefficients, it leads to a more global view on the matrix of coefficients. At the same time, we want to come closer to the ideal coefficient selection. As becomes clear from the subsequent sections, both objectives are reflected by the Bayesian model that we use: the geometrical structure is described in the prior model, whereas the conditional model explicitly appeals to the idea of an oracle's decision.

6.2.2 Plugging the threshold procedure into a fully random model

Whereas typical threshold algorithms are based on this heuristical approach that the largest coefficients capture the essential image features, *Bayesian* methods start from a full model for wavelet coefficients of the following type:

$$W = V + N$$

This is an additive model for wavelet coefficients where N is the noise vector, V is the uncorrupted wavelet coefficient vector, and W is the input (empirical) wavelet coefficient vector. Both the noise and the noise-free data are viewed as realizations of a probability distribution. We now describe how threshold procedures fit into this model.

A wavelet threshold algorithm consists of three steps: first, the observational data are transformed into empirical wavelet coefficients. The next step is a manipulation of the coefficients and finally, an inverse transform of the modified coefficients yields the result. When extending this thresholding with a Bayesian approach, we leave the three steps intact, but we build in more uncertainties in the second step. As explained in Section 2.8, the selection criterion used in the second step is based on a measure of *regularity*. This measure of significance M is a function of the observation:

$$M = m(W),$$

Wavelet coefficients with a significance below a threshold λ, are classified as noisy. With each wavelet coefficient W_s, the algorithm associates a 'label' or 'mask' variable X_s such that:

$$X_s = \begin{cases} 0, & \text{if } W_s \text{ is noisy according} \\ & \text{to the criterion, i.e. if } M_s < \lambda \\ 1, & \text{then } W_s \text{ is sufficiently clean, i.e. if} \\ & M_s \geq \lambda \end{cases} \tag{6.1}$$

In these and following equations, s represents the 'multidimensional' index of a wavelet coefficient on a given resolution level j and for a given component (vertical, horizontal, or diagonal): $s = (k, l)$. To avoid overloaded notations we omit the resolution level j and the component m in our equations, and use the simple index s. So, if no confusion is possible, we write W_s instead of $W_{j;s}^m$ or $W_{j;k,l}^m$. This classification is followed by the modification step: If $W_{\lambda s}$ denotes the modified coefficient, with subscript λ referring to the threshold value, we write:

$$W_{\lambda s} = h(W_s, M_s, X_s),$$

for some *action* $h(W_s, M_s, X_s)$. The classic *hard-threshold* procedure corresponds to

$$h(W_s, M_s, X_s) = W_s X_s.$$

It keeps the 'uncorrupted' coefficients and replaces the noisy ones by zero.

6.2.3 Threshold mask images

Figure 6.5 visualizes the binary label image X, i.e. it shows in black the position of the selected wavelet coefficients for the horizontal subband at the one but finest resolution level from the non-decimated wavelet transform of the noisy image in Figure 5.17. The one but finest scale in the wavelet transform is two scales below the original image resolution, and, as before, we use the variation on the CDF-(spline)-filters "with less dissimilar lengths" [49, 18]. Primal and dual wavelets have four vanishing moments. The mask on the left-hand side was obtained by soft-thresholding using the minimum MSE threshold. The mask on the right-hand side is obtained by a generalized cross validation threshold. Applying soft-thresholding using this mask (and its analogues for other components and scales) leads to the output in Figure 5.19.

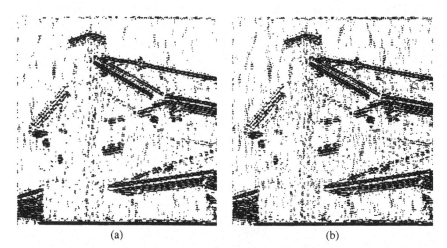

(a)   (b)

Figure 6.5. Mask or label images, corresponding to the horizontal component of the one but finest scale. Black pixels represent coefficients with magnitude above the threshold. Left: using the minimum MSE threshold. Right: using a GCV estimate of this threshold.

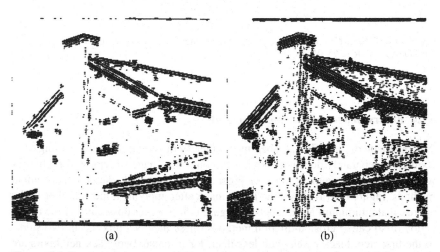

(a)   (b)

Figure 6.6. Same mask images as in Figure 6.5, here based on noise-free coefficients. Left: black pixels indicate noise-free coefficients with magnitude above the previous threshold. Right: black pixels indicate noise-free coefficients with magnitude larger than noise deviation. This corresponds to the ideal wavelet selection: if an "oracle" [76] tells us whether or not a coefficient is dominated by noise, this is the best thing we can do.

If we apply the same threshold to the noise-free coefficients, we get the left picture in Figure 6.6. We see that many of the isolated pixels have disappeared: they were due to noise. Applying the optimal selection, inspired by the 'oracle' information leads to an even more structured image in Figure 6.6(b). To conclude this discussion, we compare the result from soft-thresholding with level- and subband-dependent GCV-threshold with the result from the oracle selection, also referred to as the optimal (clairvoyant) diagonal projection. Only the three finest scales were processed. Signal-to-noise ratio is respectively 17.86 dB and 21.05 dB. The GCV result is in Figure 5.19, while Figure 6.7 contains the output from the ideal selection.

Figure 6.7. Output from the optimal (clairvoyant) diagonal projection, applied to three resolution levels. SNR = 21.05 dB.

6.2.4 Binary image enhancement methods

A comparison of the different label images clearly illustrates that thresholding each coefficient separately does not take into account the image structures. An obvious way to recover the optimal mask of Figure 6.6 (b), is trying classic enhancement methods. Figure 6.8 (a) shows the label image after applying a median filter to the approximate minimum MSE labels in Figure 6.5. Another possibility is the application of so called erosion-dilation methods: these methods proceed in two steps: in the first step, black pixels with less than, for instance, two black neighbors are removed. This erosion can be repeated several times, before going to the dilation. This second step tries to restore the eroded objects by turning white background pixels into black object pixels, if there is already an object in the neighborhood. (This neighborhood is typically a 3×3 box containing the actual pixel in its center.) Figure 6.8 (b) contains the result of this operation. It is hard to preserve the fine edge structures, while removing the noisy pixels. These operations act on the label

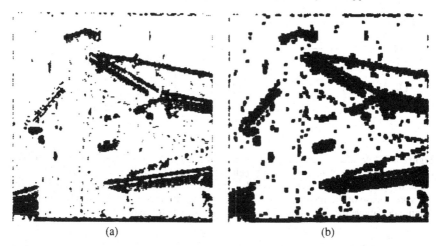

(a)   (b)

Figure 6.8. Result of elementary binary image enhancement methods on the approximate minimum MSE label image in Figure 6.5. Left: a median filter; Right: an erosion-dilation procedure.

images only and forget about the background behind them: these pixels come from a wavelet coefficient classification. We would prefer a method that can deal with the geometry *and* the local coefficient values at the *same* time. Bayes' rule tells us how we can do so.

6.2.5 Bayesian classification

The classification (6.1) in a threshold algorithm is a deterministic function of the empirical coefficients: thresholding on magnitudes corresponds to a simple step function, as illustrated in Figure 6.9. Recall that, in this text, the measure of significance is the coefficient magnitude: $M = |W|$. However, it would be interesting to examine measures based on interscale-correlations: e.g. $M_{j;s}^m = \prod_i W_{j+i;s}^m$, where j is the current resolution level and m the orientation ($m =$ HOR, VER, DIAG).

Because we want to take the spatial configurations into account, we give up this tight relation between a coefficient value and its classification. We introduce a *prior model* for coefficient classification configurations X. This prior should express the belief that clusters of noise-free coefficients have a higher probability than configurations with many isolated labels. In particular, edge-shaped clusters should be promoted. The prior model rests on the *classification* of the coefficients, not on the uncorrupted coefficients themselves. A similar idea is the use of Hidden Markov Models [44, 57, 122, 22, 27].

Next, the *conditional model* (likelihood function) states that if the classification label for a coefficient equals one, this coefficient is *probably* above the threshold. A zero label means that the corresponding coefficient is probably small. The classical, deterministic approach can be seen as an extreme case of this probability model,

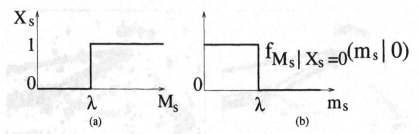

Figure 6.9. Left: The deterministic classification function for coefficient magnitude thresholding: if a coefficient magnitude M is below the threshold value λ, it is classified as noisy ($X = 0$), otherwise it is called sufficiently clean ($X = 1$). Right: this deterministic approach is a special case of the Bayesian model, where the conditional density is zero for coefficient magnitudes above the threshold if $X = 0$ and beneath the threshold if $X = 1$.

where, for example, a label $X = 0$ tells that the coefficient is *certainly* in the range $[-\lambda, \lambda]$. This appears in Figure 6.9(b).

If we have a prior distribution $P(X = x)$ and a conditional model $f_{M|X}(m|x)$, then Bayes' rule allows to compute the *posterior* probability:

$$P(X = x|M = m) = \frac{P(X = x)f_{M|X}(m|x)}{f_M(m)}$$

In a given experiment, where m is fixed, the denominator is a constant. As we explain later, it is sufficient to know the *relative* probabilities of configurations, and therefore we write:

$$P(X = x|M = m) = C \cdot P(X = x)f_{M|X}(m|x)$$

6.3 Prior and conditional model

6.3.1 The prior model

As explained above, we are looking for a multivariate model for a binary image X. Expressing a probability function for all 2^{NM} possible values in a N by M label image may be cumbersome. We therefore construct the model starting from local descriptions of clustering. The prior probability function can always be written as:

$$P(X = x) = \frac{1}{Z} \exp[-H(x)], \qquad (6.2)$$

with the *partition function Z* :

$$Z = \sum_x \exp[-H(x)].$$

$H(x)$ is the *energy function* of a configuration x: the lower the energy, the higher the prior probability. To express that this energy comes from *local* interactions only, we first define for each pixel index s in the lattice S its set of neighbors, i.e. the set of indices $\partial s \subset S$ that interact with s. We assume that $s \notin \partial s$ and $s \in \partial t \Leftrightarrow t \in \partial s$. A set of indices that are all neighbors to each other is called a *clique*. The set of all cliques is

$$\mathcal{C} = \{C \in 2^S \mid \forall s, t \in C : t \in \partial s\}$$

If the total energy of a configuration equals the sum of its clique potential functions:

$$H(x) = \tau \sum_{C \in \mathcal{C}} U_C(x_C),$$

the probability function $P(X = x)$ is named a *Gibbs distribution* with respect to the given neighborhood system $\{S, \partial\}$, after the American theoretical physicist and chemist, Josiah Gibbs, who used this model in statistical mechanics [86]. The hyperparameter τ measures the *rigidity* of the configuration. The higher τ, the less likely are status changes due to noise. As the equation indicates, the clique potential U_C only depends on the values of x in the sites that belong to C.

Whereas Gibbs distributions are based on local energies, Markov Random Fields (MRF) are based on local statistical dependencies. A Markov Random Field, relative to a neighborhood system $\{S, \partial\}$ is a probability function P with the two-dimensional Markov property:

$$P(X_s = x_s | X_{S \setminus \{s\}} = x_{S \setminus \{s\}}) = P(X_s = x_s | X_{\partial s} = x_{\partial s}).$$

This definition does not use the notion of *clique*.

The Hammersly-Clifford theorem states that MRF's and Gibbs distributions are the same:

Theorem 6.1 (Hammersly-Clifford) *A probability function is a Markov Random Field with respect to a neighborhood system if and only if it is a Gibbs distribution with respect to same neighborhood system.*

Proofs are in [25, 179]. This theorem facilitates the computation of conditional probabilities in the lattice of a Gibbs distribution: this computation only uses local information. The computation of marginal probabilities of a MRF is well served by the Gibbs distribution property. Especially in expressions like

$$\frac{P_X(u)}{P_X(v)}$$

where u and v differ in a couple of lattice points s only, it is easy to use the theorem and limit the calculations to the potentials of cliques that contain one of these points s:

$$\frac{P_X(u)}{P_X(v)} = \exp\left(-\tau \sum_s \sum_{C \in \mathcal{C}, s \in C} U_C(u_C) - U_C(v_C)\right).$$

Actually, a Gibbs distribution is mostly the only practically possible specification of a Markov Random Field: it is hard to check whether a collection of local conditional probabilities form a coherent set for a Markov Random Field [85].

The application of this MRF's and Gibbs distributions in image analysis and image processing is still growing and our list [85, 100, 169, 179] is nothing but a snapshot. An often used Gibbs distribution is the Ising model, named after the German physicist Ising, who used it to explain ferromagnetism [102]. The neighbors of a pixel with index s is a 3×3 submatrix, excluding s in its center. The model only considers pairs of neighbors. Other cliques in the system have no energy. The total energy is:

$$H(x) = \sum_{\{s,t\} \in \mathcal{C}} x_s x_t.$$

6.3.2 The conditional model

We also need a conditional density $f_{M|X}(m|x)$. Whereas the prior describes the clustering of significant wavelet coefficients, this conditional model deals with the *local* significance measure. Therefore we write

$$f_{M|X}(m|x) = \prod_{s \in S} f_{M_s|X_s}(m_s|x_s).$$

This density expresses that if the label $X_s = 1$, i.e. if the corresponding wavelet coefficient is sufficiently noise-free, a large value of M_s is probable. Referring to the ideal selection procedure, we now impose the idea that selected coefficients should have an untouched value above the noise deviation σ. This means that if $X_s = 1$, V_s is for instance uniformly distributed on $[-\mu, -\sigma] \cup [\sigma, \mu]$, μ being the maximum coefficient magnitude, which is a parameter of the model that has to be determined. If the noise $N_s \sim n(0, \sigma)$ has a Gaussian density, it is easy to verify that

$$f_{W_s|X_s}(w|1) = \frac{1}{2(\mu - \sigma)} \left[\Phi(w + \mu) - \Phi(w + \sigma) + \Phi(w - \sigma) - \Phi(w - \mu) \right],$$

where $\Phi(z)$ is the cumulative Gaussian distribution. A similar argument leads to the following conditional model for coefficients dominated by noise:

$$f_{W_s|X_s}(w|0) = \frac{1}{2\sigma} \left[\Phi(w + \sigma) - \Phi(w - \sigma) \right].$$

Figure 6.10 shows these density functions for $\sigma = 1$ and $\mu = 20$. This model expresses that if a label $X_s = 0$, the corresponding coefficient *probably* has a small magnitude, but magnitude is no longer a strict selection criterion: a small coefficient might be important and a large coefficient might be replaced by zero.

Other, perhaps more realistic models follow from the assumption that the important, noise-free coefficients are exponentially distributed:

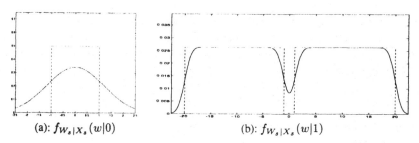

(a): $f_{W_s|X_s}(w|0)$ (b): $f_{W_s|X_s}(w|1)$

Figure 6.10. Conditional probability densities in Bayesian model. The model expresses that if a label $X_s = 0$, the corresponding coefficient *probably* has a small magnitude, but magnitude is no longer a strict selection criterion: a small coefficient might be important and a large coefficient might be replaced by zero.

$$f_{V_s|X_s}(v|1) = \frac{\rho e^{\rho\sigma}}{2} e^{-\rho|v|}.$$

This leads to the following expression for the coefficients corrupted by noise:

$$f_{W_s|X_s}(w|1) = \frac{\rho\, e^{\sigma\rho + \sigma^2\rho^2/2}}{2}$$
$$\left[e^{-\rho w}\Phi(w - \sigma - \sigma^2\rho) + e^{\rho w}(1 - \Phi(w + \sigma + \sigma^2\rho)) \right].$$

Even more general models for noise-free wavelet coefficients are Laplacian distributions [154]:

$$f_V(v) = K\, e^{-|kv|^{\kappa}}.$$

Typical values for κ range between 0.5 and 0.8.

6.4 The Bayesian algorithm

6.4.1 Posterior probabilities

From Bayes' rule, we can compute the posterior probabilities

$$P(X = x|M = m) = \frac{P(X = x)f_{M|X}(m|x)}{f_M(m)}$$

With these probabilities, a *Bayesian decision rule* leads to an estimation of the optimal label. The Maximum A Posteriori (MAP) procedure chooses the mask x with the highest posterior probability. The Maximal Marginal Posterior (MMP) rule is a more local approach: it computes in each site s the marginal probabilities:

$$P(X_s = 1|M = m) = \sum_{x} x_s P(X = x|M = m),$$

and if this probability is more than 0.5, the pixel gets value 1. Both decision rules have a binary outcome: each coefficient is classified as noisy ($X = 0$) or relatively uncorrupted ($X = 1$). We would like to exploit the entire posterior probability: the posterior mean value

$$E(X_s|M = m) = P(X_s = 1|M = m)$$

preserves all information. It is a minimum least squares estimator. This classification leads to a posterior 'expected action':

$$E(W_{\lambda s}|M) = h(W_s, m_s(W), 1)P(X_s = 1|M)$$
$$+h(W_s, m_s(W), 0)P(X_s = 0|M).$$

If $h(W_s, M_s, X_s) = X_s W_s$, this is:

$$E(W_{\lambda s}|M) = W_s E(X_s|M = m) = W_s P(X_s = 1|M).$$

Unlike most thresholding methods, this is not a binary procedure: using the posterior probability leads to a more continuous approach.

6.4.2 Stochastic sampling

The computation of $P(X_s = 1|W)$ involves the probability of all possible configurations x. Because of the enormous number of configurations, this is an intractable task. The sum we have to compute is of the following form:

$$\mu_s = \sum_x f_s(x)P(X = x|M = m)$$

where in this case $f_s(x) = x_s$ and $\mu_s = P(X_s = 1|M = m)$.

To estimate this type of sum (or integral for random variables on a continuous line), one typically uses stochastic samplers. These methods generate subsequent samples $X^{(i)}$, not selected uniformly, but in proportion to their probability. This allows to approximate the matrix of required marginal probabilities by the mean value of the generated masks:

$$\hat{\mu}_s = \sum_i X_s^{(i)}.$$

Mostly, the samples are generated, not independently of each other, but in a chain, hence the name Markov Chain Monte Carlo (MCMC) estimation. The next sample is generated, starting from the previous one. One advantage of this procedure is that knowledge of the *relative* probabilities of the candidates is sufficient. The probability ratio of two subsequent samples:

$$r^{(i)} = \frac{P(X^{(i+1)}|M)}{P(X^{(i)}|M)}$$

is the only quantity needed by the algorithm, and if

$$P(\boldsymbol{X} = \boldsymbol{x}|M) = \frac{1}{Z f_{\mathbf{M}}(\mathbf{m})} \exp[-H(\boldsymbol{x})] f_{\mathbf{M}|\mathbf{X}}(\boldsymbol{m}|\boldsymbol{x}),$$

there is no need for the enormous computation of the partition function $Z f_{\mathbf{M}}(\mathbf{m})$.

We use the classic Metropolis MCMC sampler [134]. The chain of states is started from an initial state $\boldsymbol{X}^{(0)}$. The successive samples $\boldsymbol{X}^{(i)}$ are then produced as follows: a candidate intermediate state is generated by a local random perturbation of the actual state. Then the probability ratio r of the actual state and its perturbation is computed. Since the Gibbs distribution is based on local potential functions, only positions s whose mask labels are switched by the perturbation or which have a switched label in their neighborhood ∂s are involved in the computation. If the candidate has a higher probability than the actual state, i.e. if the probability ratio is larger than one, then the new state is accepted, otherwise it is accepted with probability equal to r. To generate a completely new sample, we repeat this local switching for all locations in the grid.

6.5 Parameter estimation

6.5.1 Parameters of the conditional model

The conditional model $f_{V_s|X_s}(v|1)$ or $f_{W_s|X_s}(w|1)$ is for instance uniform or exponential. This model contains a hyperparameter. It is not so hard to fill in this parameter using the observed, noisy wavelet coefficients. In our approach, we mostly use the uniform model on $[\sigma, \mu]$ for which it is easy to prove that the expected highest magnitude $E|V|_{\max}$ equals:

$$E|V|_{\max} = \frac{N\mu + \sigma}{N + 1}.$$

A good measure for the noise variance is the average energy removed by the minimum MSE-threshold:

$$\hat{\sigma}^2 = \frac{1}{N} \sum_{i=1}^{N} (W_{\lambda i} - W_i)^2.$$

Since the influence of the noise on the largest coefficients is relatively small, we take:

$$\hat{\mu} = \frac{(N + 1)|W|_{\max} - \hat{\sigma}}{N}.$$

6.5.2 Full Bayes or empirical Bayes

The prior energy model contains a parameter τ:

$$H(x) = \tau \sum_{C \in \mathcal{C}} U_C(x_C)$$

It determines the local rigidity of the prior. The higher its value, the larger the energy difference between the two states of a given pixel in the label image. The choice $\tau = 0$, for instance, disregards spatial structures.

To find a good value for this parameter, there exists at least two approaches. The *fully* Bayesian approach considers this parameter as an instance of still another density and assigns a prior distribution f_τ to τ. The posterior density for this parameter is then:

$$f_{\tau|X}(\tau|x) \propto f_\tau(\tau) P(X = x|\tau).$$

The posterior probability of the full set of unknowns X, and τ, given the observation $M = m$ then satisfies:

$$P(X = x, \tau|M = m) \propto f_{M|X}(m|x) P(X = x|\tau) f_\tau(\tau).$$

From these expressions, we can — for instance — find values for M and τ with maximum posterior probability.

The *empirical* Bayes approach maximizes the likelihood of the actual label image x:

$$L(\tau) = P(X = x),$$

where the probability function on the right hand side depends on the rigidity τ through the energy function $H(x)$. This maximum likelihood estimation (MLE) has two practical problems. First, the computation of the likelihood function is extremely hard, due to the intractable partition function in (6.2). Therefore, and since the rigidity parameter controls the local behavior of the label image, we use a pseudo-likelihood method: we maximize the product of "local" likelihood functions:

$$PL(\tau) = \prod_{s \in S} P(X_s = x_s | X_{\partial s} = x_{\partial s}, \tau).$$

This maximum pseudo likelihood estimation (MPLE) leaves us with the second problem: we have no real instance of the probability function $P(X = x)$, because we only have *noisy* measurements M. The probability function of X supposes that X is the optimal selection of wavelet coefficients. This selection is based on the uncorrupted values V being above or below σ, which we do not know. A typical solution for this kind of incomplete observation could be the EM-algorithm, also known as the Braum-Welch algorithm [57, 66] This is an iterative procedure, alternating between two steps which could be labeled as (the computation of the)

Expectation (or E-step) and Minimization (or M-step), hence the abbreviation EM. Although this idea is generally applicable, the elaboration is different for all models.

Alternatively, we can simply assume that the *local* behavior of the mask obtained by thresholding the *noisy* coefficients approaches the rigidity of the optimal selection. The choice of the threshold is of course crucial in this approximation: we cannot take $\lambda = \sigma$, pretending $W \approx V$, since this would generate highly noisy masks, with little structure from the optimal selection. A mask generated by the minimum MSE or GCV is generally still too noisy, as becomes clear from a comparison of the labels in Figure 6.5 with the ideal one in Figure 6.6(b). This can be helped by applying a median filter to the minimum GCV labels, as in Figure 6.8. As mentioned before, this median filter does not take into account the background of the individual labels, like the conditional density in a Bayesian approach. Therefore it is less appropriate for the actual correction of the label images, but it may do a good job in estimating the rigidity factor τ of the optimal selection mask on a local basis. Another possibility is the universal threshold: this threshold eliminates all noise with high probability, at the risk of loosing parts of the underlying structure.

6.6 The algorithm and its results

6.6.1 Algorithm overview

This is a schematic overview of the subsequent steps of the algorithm:

1. Compute the non-decimated wavelet transform W of the input.
2. At each level and for each component, select the appropriate threshold. This threshold generates an initial label image $X^{(0)}$.
3. Apply a median filter to $X^{(0)}$ and estimate the prior parameter τ from the result, using a maximum pseudo-likelihood estimator.
4. Run a stochastic sampler to estimate for each coefficient at the given resolution level the probability $P(X_s|W)$. Use $X^{(0)}$ from the previous step as the starting sample. A Markov Chain Monte Carlo algorithm produces the sequence of samples.
5. $\hat{W}_{\lambda s} \leftarrow W_s P(X_s = 1|W)$.
6. Inverse wavelet transform yields the result.

6.6.2 Results and discussion

We now apply the procedure to the image with artificial noise in Figure 5.17. Figure 6.11(a) shows the mask image after ten MCMC-iterations. To be more correct: this image represents for each coefficient the posterior probability $P(X_s = 1|W)$ of its label being one. More iterations (up to 100) did not improve the output quality. This output appears in Figure 6.11(b). Signal-to-noise ratio is 18.69 dB. Looking at the posterior probabilities, and comparing this with the objective mask in Figure 6.6(b), we see that most spurious labels from the threshold procedure indeed have a low posterior probability. The important structures that are present in the label image corresponding to the MSE-threshold, are preserved: the coefficients belonging to these structures have high probabilities. Nevertheless, it seems to be hard to recover clusters of small coefficients, even if these structures appear in the optimal selection.

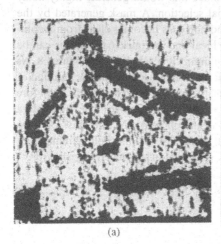 

(a)   (b)

Figure 6.11. Left: label image for the wavelet coefficients of the image in Figure 5.17 after ten MCMC iterations. Consequently, this image has 11 grey values. A pixel value is an estimate of the marginal posterior probability $P(X_s = 1|M)$. Right: the algorithm output. Three resolution levels were processed. Signal-to-noise ratio is 18.69 dB.

We also illustrate the method with the 'realistic" MRI-image of a knee in Figure 5.25. Figure 6.12(a) has the output of the Bayesian algorithm, applied to the first and second resolution level. Figure 6.12(b) shows the label image for the vertical subband at the finest resolution level, to be compared with the selection of a minimum GCV-threshold, depicted in Figure 6.12(c). The latter selection is based on local regularity (magnitude) and shows far less geometrical structure.

6.6.3 Related methods

Our prior model was designed to describe geometrical correlations among coefficients within a given subband (scale and component). This type of correlation typically appears in two-dimensional wavelet transforms, especially in image analysis. Interscale correlations, present in all dimensions, are not captured by our prior model, although this is possible, as in [57].

Another difference is the meaning of the label values X_s, and, consequently the design of the conditional model. Unlike the labels in [46, 57], a label one in our algorithm means that the corresponding noise-free coefficient is certainly larger than

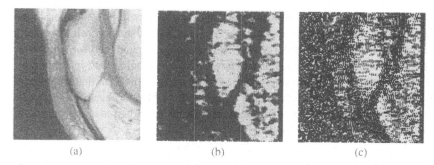

(a)   (b)   (c)

Figure 6.12. (a) Output of the Bayesian algorithm, applied to the first and second resolution level of the image in Figure 5.25. (b) and (c): Selection masks for vertical subband at the finest resolution level. The image in (b) has eleven grey levels, it represents for each coefficient an MCMC-estimate of the posterior probabilities of being important. The MCMC procedure used ten iterations, hence eleven grey levels, from zero to one. The last image is binary: black pixels correspond to coefficients that are preserved by a minimum GCV-threshold. This selection is based on local regularity (magnitude) and shows far less geometrical structure.

σ. The conditional model is explicitly inspired by the idea of finding the optimal diagonal projection of [76]. We do not compute a posterior mean $E(V_s|W)$, but rather a posterior expected action: $E(W_{\lambda s}|W)$.

This algorithm was inspired by previous work by Malfait et al. [125, 124], although their algorithm is based on Hölder regularity, and therefore looks at the evolution of coefficients through scales. Our algorithm uses coefficient magnitudes at one scale only, because this leads to more stable computations. Second, unlike the work by Malfait et al. the algorithm described in this text aims at the optimal coefficient selection, and the conditional model has been designed with this objective in mind. Third, all model parameters in our algorithm are determined automatically, in an empirical or heuristical way: there is no need for learning, the algorithm adapts itself to a given image.

6.6.4 Possible extensions

We introduced a geometrical prior model for configurations of important two-dimensional wavelet coefficients. This model was combined with a threshold algorithm for noise reduction. Instead of just using coefficient magnitudes, one could involve interscale correlations. One possibility are multiscale Markov Random Fields [57], where the interlevel correlation appears in the prior model. Alternatively one could build interlevel correlations into the measure of regularity to be used in the conditional model. To our knowledge, this possibility has not been investigated so far. This approach relies on a correlation driven deterministic algorithm, like [180] and it should be faster than a multiscale prior model in a Bayesian procedure.

Further experiments with other priors and different models for noise-free coefficients, like Laplacian distributions, are other possible extensions.

While looking for more sophisticated models, one should pay attention to the algorithm complexity: this is a crucial factor in this Bayesian approach. Especially the parameter estimation by the MPLE method could be examined for simplification. This estimation is already much faster than a full MLE, but yet requires a lot of computation. It is also interesting to reconsider the problem that we observe noisy masks: hence, we cannot be sure of the rigidity of the ideal selection mask. Is there any method to estimate this rigidity, taking this perturbation into account, for instance using the EM-philosophy?

6.7 Summary and conclusions

This chapter has investigated the possibilities of a Bayesian procedure to improve the results of a wavelet thresholding procedure. This procedure was designed for application in image noise reduction and it combines two objectives:

1. We want to capture the correlations in wavelet coefficients due to edge singularities. This type of singularities is specific for more-dimensional data, like images. The *prior* model in our procedure takes these line singularities into account: the model is based on *geometrical* properties: it favors clusters of important coefficients.
2. With the aid of this geometrical prior, we aim at mimicking the optimal coefficient selection. This is reflected in the *conditional* model.

The algorithm succeeds in finding more structure in the coefficient selection, which results in an output with better preserved edges. It would be interesting to quantify this gain in contrast. A more sophisticated conditional model, based on Laplacian distributions for uncorrupted wavelet coefficients, as well as the never ending search for good prior models are other topics for further research.

7

Smoothing non-equidistantly spaced data using second generation wavelets and thresholding

A classical (first generation) wavelet transform assumes the input to be a regularly sampled signal. In most applications of digital signal processing or digital image processing, this assumption corresponds to reality. In many other applications however, data are not available on a regular grid, but rather as non-equidistant samples. Examples in this chapter illustrate what happens if we use classical wavelet transforms, pretending that the data are equispaced: the irregularity of the grid is reflected in the output.

Working with wavelets on irregular grids guarantees a smooth reconstruction [61]. This chapter investigates closeness of fit: it turns out that stability issues make it hard to hit this target from the wavelet domain. A close fit in terms of wavelet coefficients may be not so good after reconstruction. As a consequence of this, the connection between coefficient magnitude and importance is not so clear anymore: omitting a small coefficient may cause an important bias after reconstruction.

Nearly all existing wavelet based regression of non-equispaced data combines a traditional equispaced algorithm for fitting with a "translation" of the input into an equispaced problem. Possible techniques to do so are:

1. Interpolation in equidistant points [96, 113, 65]
2. Projection of the equispaced result onto the irregular grid [33, 15, 152, 147, 17, 16]. Some of these methods pay special attention to the approximation of the scaling basis and the projection coefficients therein.

In general, preprocessing the data changes their statistics. White noise, for instance, becomes colored, or stationarity may be lost.

The second generation wavelets approach for de-noising non-equidistant samples is new. A discussion of the unbalanced Haar transform [87] for regression appears in [64], which also contains an excellent overview of first generation approaches. The origin of the stability problems when using second generation wavelets is not yet fully understood. This chapter proposes some possible explanations.

7.1 Thresholding second generation coefficients

7.1.1 The model and procedure

For this text, we suppose that the data live on a fixed, irregular grid:

$$y_i = f(x_i) + \eta_i, \quad i = 1, \dots, N.$$

Some methods based on classical wavelets use a 'preconditioning' of the data (by interpolation in equidistant points, for instance). In that case, it may make a difference starting from a random model for the data points:

$$y_i = f(X_i) + \eta_i, \quad i = 1, \dots, N,$$

where the X_i come from a random distribution. In our approach, however, there is no need for specifying how the data points were selected.

We just apply a second generation wavelet transform to the input. Since this transform takes into account the lattice of the data, the noise standard deviation is different for each coefficient, even if the noise on the input data had a constant standard deviation. This lack of homoscedasticity makes thresholding difficult: if the amount of noise is different for each coefficient, it is hard to remove it decently by only one threshold.

Nevertheless, if we know the covariance structure Q of the input noise, we can compute the variance fluctuation in the wavelet domain, as in (2.20):

$$S = \tilde{W} Q \tilde{W}^T$$

If Q is a banded matrix, S can be computed in a linear amount of time. In practical cases, the exact values of Q are often unknown, but the structure of Q may be known, i.e. Q may be known up to a constant. The case of stationary white noise, for instance, corresponds to: $Q = \sigma^2 I$, with I the identity matrix.

The normalised coefficients

$$\tilde{w}_i = w_i / \sqrt{S_{ii}}$$

do have a constant variance and thresholding these coefficients makes more sense than thresholding the original ones.

7.1.2 Threshold selection

The matrix Q may not contain the exact variances, but only the structure of the covariance matrix. This is the case if we know the structure of the correlation of the input noise. In many practical situations, for instance, it is reasonable to assume that the noise is white and stationary without specifying the exact noise level.

To find a good threshold without using an estimate for this noise level, we rely on the GCV-procedure of Chapter 4:

$$GCV(\lambda) = \frac{\frac{1}{N}\|\tilde{w} - \tilde{w}_\lambda\|^2}{[\frac{N_0}{N}]^2}.$$

In this equation $\tilde{w}_\lambda$ is the vector of thresholded normalized coefficients and, as usual, N_0/N stands for the fraction of coefficients replaced by zero by this particular threshold value λ.

7.1.3 Two examples

We illustrate the effect of using second generation wavelets with two examples. In the first example, the grid was obtained by selecting $N = 2048$ points x_k at random between 0 and 1. These points were ordered and used as sampling points for the "heavisine" function [78]:

$$\begin{aligned} f(x) &= 4\sin(4\pi x) - \text{sign}(x - 0.3) - \text{sign}(0.72 - x) \\ y_k &= f(x_k) + \eta_k, k = 1, \dots, N \end{aligned}$$

The figures 7.1 show a detail of 600 points. The algorithm used a lifting scheme, based on cubic interpolation for prediction and a two-taps update filter. We can neglect the grid structure, i.e. we run the algorithm with an equidistant grid, this means that we are smoothing the data (k, y_k) instead of (x_k, y_k). The result is noisy, because the regular grid transform does not correspond to the real grid. The spikes in the result are inherent for this simple threshold algorithm. More sophisticated algorithms should be able to remove them. The curve in between these spikes however is much smoother if we use second generation wavelets.

A second example is a damped sine ($f(x) = e^{-x}\sin 4\pi x$) on an extremely irregular grid. This grid was constructed as follows: we choose approximately 100 samples at random between 0 and 0.2, about 10 samples between 0.2 and 0.4 and about 1940 samples between 0.4 and 1. Figure 7.2 plots the grid point versus the point number. If we add white and stationary noise to this function, we get the upper plot of Figure 7.3. The left part of this plot looks less noisy, but this is because data points in the right tail are much closer to each other. As for the previous example, second generation wavelets give a generally smoother result, but in this case, this scheme introduces a tremendous bias, not only in the region with few data points, but also at places where data are given close to each other. One could argue that this example is somehow artificial. Moreover, the phenomena seem to appear mostly at coarse scale, and it is a common practice to leave coefficients at coarse scales untouched. Nevertheless, if we run the same algorithm pretending the grid to be regular, the result is quite fair, apart from the grid irregularities, of course. We now investigate where this bias comes from and what we can do to make the second generation algorithm perform at least as well as the "first" generation wavelets.

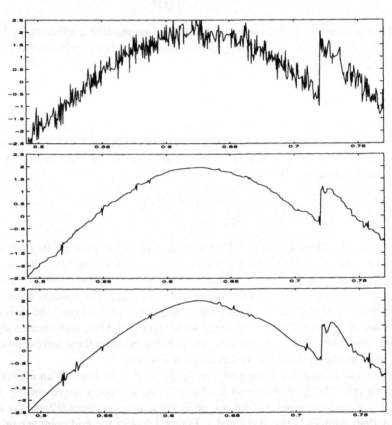

Figure 7.1. Example 1: Top: noisy 'HeaviSine' function on a "not too" irregular grid. The grid was obtained as an ordered set of uniformly chosen points on the interval [0, 1]. Middle: result of a threshold algorithm on a classical wavelet transform. We run the lifting scheme but tell the algorithm that the grid is regular. The result is noisy, because the regular grid transform does not correspond to the real grid. Bottom: result of the same algorithm based on the actual grid. In both cases, we use GCV to estimate the MSE-threshold.

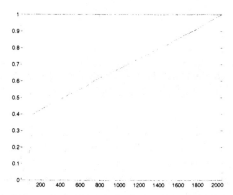

Figure 7.2. The grid of Example 2: this very irregular grid was constructed as follows: we choose approximately 100 samples at random between 0 and 0.2, about 10 samples between 0.2 and 0.4 and about 1940 samples between 0.4 and 1.

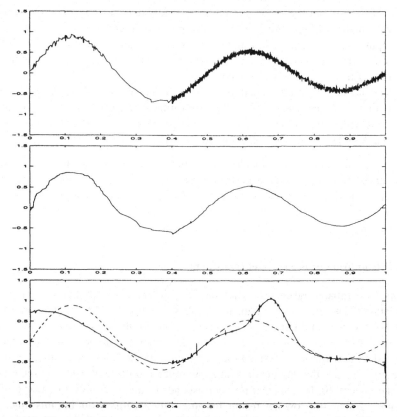

Figure 7.3. Example 2: Top: noisy signal ($f(x) = e^{-x} \sin 4\pi x$) on the grid of Figure 7.2. Middle: result of a threshold algorithm on a classical wavelet transform. The lack of smoothness in this result reflects the irregularity of the grid. Using second generation wavelets leads to a much smoother result, but for this example, this scheme causes an unacceptable bias.

7.2 The bias

7.2.1 The problem

The bias comes from the fact that the second generation wavelet transform may be far from orthogonal. This appears in several effects, which sometimes enhance each other.

1. A small coefficient may have a wide *impact*, especially when it is related to a region with only a few samples. Thresholding it causes an important effect in the original domain.
2. Basis functions sometimes have a large overlap, especially in the neighborhood of boundaries. Large individual coefficients may then compensate each other, resulting in a signal with a relatively small energy. Thresholding these coefficients destroys the balance between the large coefficients and causes artifacts: hidden components suddenly become visible.
3. A transform has a bad condition number if it is sensible to errors on the input. As a matter of fact, thresholding can be considered as an artificial error on the input, and typically, the threshold is much larger than machine-precision! If the transform is not stable, there is no guarantee that the output is close to the input, even if it is so in wavelet-domain.
4. The threshold is proportional to the standard deviation of a coefficient. Unlike in the stable case, coefficients with large variance may correspond to basis functions with large energy. Or, equivalently, dividing wavelet coefficients by their standard deviation may cause important coefficients to become relatively small.

The bad condition of such a wavelet transform plays a role in other applications too, of course. From the statistical point of view, we are specifically interested in the interaction between variance normalization and bias, as described in the last item of this enumeration. In short, a bad conditioned transform makes it difficult, if not impossible, to predict the effect of a threshold on a coefficient.

7.2.2 Condition of the wavelet transform

Table 7.1 compares the condition number for different wavelet transforms on different lattices. The first row contains the condition numbers of first generation wavelets, treating the boundaries by periodic extension. For the second row, we still have an equidistant grid, but now the boundaries are processed in the second generation way. The third row was obtained for a transform on the "close-to-regular" grid, as in Figure 7.1, i.e. the input is defined on an ordered set of uniformly chosen points on the interval $[0, 1]$. The next row corresponds to the extremely irregular grid in Figure 7.2. The last row has results for an evenly irregular grid, but this time, the zone with little data is right in the middle of the interval, instead of at the left side. In this example, there should be less interaction of the sparse data zone with the boundary. All transforms use $N = 2048$ data points and $\tilde{n} = 2$ vanishing moments for the primal wavelet function, which corresponds to a two-taps update filter in the lifting scheme. The prediction is a linear interpolation in the first column and a cubic interpolation in the second one.

	linear interpolation	cubic interpolation
(1)	.754E1	.607E1
(2)	.194E3	.755E2
(3)	.104E4	.247E5
(4)	.400E4	.253E6
(5)	.272E3	.172E6

Table 7.1. Condition numbers for wavelet transforms on different lattices: (1) equidistant grid, periodic extension, (2) equidistant grid, 2nd generation, (3) uniformly random samples, (4) lattice of Figure 7.2, (5) an evenly irregular lattice as (4) but this time, the sparse zone is in the middle of the interval. The notation $q\mathrm{E}a$ stands for $q \times 10^{a}$.

To eliminate all normalization effects, and to concentrate on the obliqueness of the basis, we consider transforms that map coefficients in a normalized scaling basis onto coefficients in a normalized wavelet basis. The according normalization at the beginning and at the end is included in the condition numbers.

7.2.3 Where does the bad condition come from?

As Table 7.1 indicates, bad condition follows from several factors and from the interaction between those:

1. The primal lifting (update) step plays an important role in this phenomenon: it turns out that an update filter with two taps $A_{j,k}$ and $B_{j,k+1}$ defines a wavelet function at scale j and place k as a combination of scaling function at two scales [164]:

$$\psi_{j,k} = \varphi_{j+1,2k+1} + A_{j,k}\varphi_{j,k} + B_{j,k+1}\varphi_{j,k+1}.$$

If the update filter coefficients are large, ψ_{jk} is close to the subspace spanned by the scaling functions $\varphi_{j,k}$ at the same scale.
2. These problems appear to have most consequences close to the boundaries of the interval, and less in the middle. Scaling functions at the boundaries tend to have heavy tails and these tails cause an important overlap of $\psi_{j,k}$ with $\varphi_{j,k}$ or $\varphi_{j,k+1}$. Most of the problems seem to arise from the fact that near the boundaries, the same set of even points is used to predict several odd points. This is because we cannot choice them symmetrically around the odd point if this odd point is close to the boundary.
3. Pretending the grid to be regular eliminates a great deal of the bias. This indicates that not only boundary problems have an impact: the irregularity itself also creates or enhances instability.

4. It is clear that the lifting theory as such neglects the notion of scale: if a sequence of dense samples is followed by a large gap, the transform operates on phenomena at different scales in one single step. In one way or another, the transform should be re-organized so that it deals with phenomena at one scale in each step. This reordering of downsampling the coefficients however does not seem so easy.

We remark that at least the Haar transform remains orthogonal on an irregular grid. For the CDF 2,2-transform, which corresponds to linear interpolation prediction and a simple update, the problems remain marginal. In both cases, there is almost no mixture of scales possible: one (for Haar) and even two (CDF 2,2) prediction points never show a structure with two different scales. In a cubic interpolation scheme, however, the four interpolation points may reflect phenomena at two different scales, for instance, if three points are close to each other and the fourth is at a long distance from this cluster.

5. Heavy tails are partly a consequence of the prediction (dual lifting) step, which determines the primal scaling functions. So, the interaction of both prediction and update seems to be responsible for at least part of the problem. Figure 7.4 illustrates that a small error in one of the interpolation points may cause a serious error in the points where this interpolating polynomial is used as a prediction. The figure shows the errors caused by a unit error in one of the interpolation points: this function is the *difference* between the correct interpolating polynomial (not shown here) and the polynomial that comes out if the error in the first interpolating point (0.01 in the example) equals one. This difference or error function itself is a Lagrange interpolating polynomial.

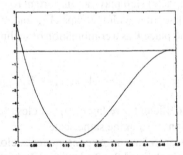

Figure 7.4. Effect on the interpolating polynomial of an error in the first interpolating point. This function is the difference between the correct interpolating polynomial (not shown here) and the polynomial that comes out if the error in the first interpolating point (0.01) equals one. This illustrates the problem that this error function may become large if the interpolating points are far from equidistant.

7.3 How to deal with the bias?

Essentially, there are two possible ways to overcome the problem of the bad condition. The first is trying to modify the transform so that it becomes more stable. In Section 7.3.3, we propose some modifications. Section 7.3.4 presents an alternative solution, using a wavelet transform on a regular grid. Inspecting this transform allows to examine which coefficients are dangerous to threshold, and to find an appropriate value for these coefficients.

7.3.1 Computing the impact of a threshold

In the first instance we try to save the coefficients that correspond to large energy basis functions from thresholding. We examine for each coefficient the influence of a threshold proportional to its noise level $\lambda = k\sigma_i$. We have that

$$\sigma_i = \sqrt{(\tilde{W}Q\tilde{W}^T)_{ii}}.$$

We assume that the input noise is uncorrelated (white) and second order stationary: $Q = \sigma^2 I$. Each coefficient w_i corresponds to a basis function. The 2-norm of this function can be computed as:

$$E_i = \int_{-\infty}^{\infty} \psi_i^2(x)\,dx = \sqrt{(W^T T_J W)_{ii}} = \sqrt{\left[(\tilde{W} T_J^{-1} \tilde{W}^T)^{-1}\right]_{ii}},$$

where $W = \tilde{W}^{-1}$ is the inverse wavelet transform matrix and T_J is a diagonal matrix containing the squared norms of the scaling function at the initial, fine resolution:

$$T_{J,kk} = \int_{-\infty}^{\infty} \varphi_{J,k}^2(x)\,dx \approx \int_{(x_{k-1}+x_k)/2}^{(x_k+x_{k+1})/2} 1^2\,dx.$$

This norm E_i is a measure for the effect of a "unit-threshold". The total effect Δy of thresholding is given by the following expression of impact:

$$\Delta y = k\sqrt{(\tilde{W}Q\tilde{W}^T)_{ii}\left[(\tilde{W} T_J^{-1} \tilde{W}^T)^{-1}\right]_{ii}}.$$

For orthonormal transforms on a regular grid and with uncorrelated, stationary noise, this effect would be independent of W: it only depends on the threshold value $\Delta y = k\sigma = \lambda$. Figure 7.5 shows the result if we preserve coefficients with a large impact from thresholding. The most serious bias has gone, but the result has lost smoothness and it is difficult to define a threshold between coefficients with large and small impact.

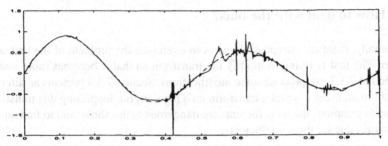

Figure 7.5. Result if we preserve coefficients with a large impact from thresholding. The most serious bias has gone, but the result has lost smoothness and it is difficult to define a threshold between coefficients with large and small impact.

7.3.2 Hidden components and correlation between coefficients

The computation in the previous section only takes into account the 2-norm of separate basis functions. The inner product of two functions, which is responsible for inter-coefficient correlations, does not appear in the algorithm. The peaks in the result are the consequence of this approach, as illustrates the following example. Figure 7.6 shows an experiment where one particular second-generation wavelet coefficient of the noisy signal was replaced by zero. Inverse transform reveals a tremendous effect. The coefficient had a rather large magnitude, and apparently also a wide impact, but comparison of the results in Figure 7.6 and Figure 7.3 indicates that the same coefficient was classified as not important by the threshold algorithm. This is because not only its magnitude was large, but so was its variance. If we remove the same threshold from the noise-free wavelet coefficients, we get the reconstruction in Figure 7.7. The difference with the original function is hardly visible. The threshold algorithm was right to remove it. A sim-

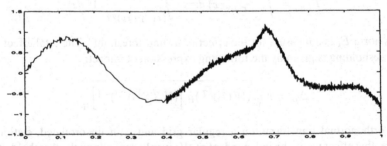

Figure 7.6. Reconstruction after removing one coefficient from the noisy transform. The effect is enormous, but the coefficient was rather big.

ple example in $\mathbb{R}^3$ makes clear what happens. Suppose we have the basis vectors $\{(-1/2, \sqrt{3}/2, 0), (-1/2, -\sqrt{3}/2, 0), (1, 0, \varepsilon)\}$. If ε is small, this basis has an extremely bad condition. Suppose the noise is $(0, 0, \varepsilon)$ in the canonical basis, then its

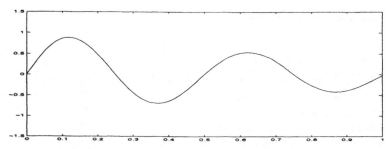

Figure 7.7. Reconstruction after removing the same coefficient as in Figure 7.6 from the noise-free transform. The effect is quasi nihil.

coordinates in this oblique basis are $(1, 1, 1)$. If one or two of these coordinates are thresholded, "hidden components" become clear. This bad condition can only be detected with a global analysis. An analysis of inner products does not suffice since none of the basis vectors is close to another one. What typically happens is this: at a certain stage in the wavelet transform, a wavelet function is very close to the some scaling functions at the same scale, as explained in Section 7.2.3. These scaling functions are further processed into a wavelet basis. The space spanned by this basis is of course still close to the first wavelet function, but the individual functions from this basis do not necessarily show large inner products with this first wavelet function.

In the example of Figure 7.6, the noise made small coefficients big, because it did not fit well into the oblique basis. Removing some of these coefficients uncovers these hidden components. The result of Figure 7.5 does not contain the same bias as in Figure 7.6. This means that the computation of the impact of the coefficients saved the coefficient of Figure 7.6 from being thresholded. This is not what we want: not only it does not correspond to what the noise-free coefficient says (this is what Donoho and Johnstone call the "oracle"), but also, if we keep this large, purely noisy coefficient, we have to keep all the others that compensate for its effect. These are hard to find, and if we find them, we end up with a result without any noise-*reduction* at this place. We would like to remove all of these large noise coefficients and therefore we want a reliable estimation of the noise-free signal: this estimation does not have to be smooth, but it should learn us which of the big coefficients are really important and which are due to noise. Unlike the classical (bi-)orthogonal transform, the second-generation transform no longer guarantees that coefficients with a large magnitude are important.

Another unpleasant consequence is the fact that scaling coefficients which are not further transformed may carry a lot of noise too. Most algorithms do not threshold scaling coefficients, and this may uncover, once more, hidden noise components. A reliable estimation of the noise-free signal could give us an idea of the effect of the noise on the low-resolution scaling coefficients.

7.3.3 Stabilizing modifications

Whatever the exact origin of the instability is, fact is that, unlike a Riesz basis, an unstable basis is far from orthogonal. We could solve stability problems by orthogonalizing the basis, for instance using Gram-Schmidt orthogonalization. This solution is computationally expensive: it has cubic complexity, which is unacceptable in the framework of a fast wavelet transform. Moreover, the basis functions after orthogonalization have infinite support. This orthogonalization therefore may serve as a benchmark for other stabilization procedures. The example of Section 7.3.2 illustrates that most problems come from interscale correlations, rather than intrascale instability. An approximate orthogonalization of subspaces W_j reduces most of the instabilities [153]. To preserve primal vanishing moments and locality, the orthogonalization is not exact. The procedure could be described as *local semi-orthogonalization*.

Another remedy concentrates on the prediction step near the boundaries. Instead of keeping the same number of prediction points, one could reduce this number near the boundaries. This avoids the same points being reused for the prediction in several odd points, since this situation is an source of stability problems. The price to pay are dual vanishing moments near the boundary.

A third possibility is a radical rearrangement of the lifting procedure. Instead of a global splitting into even and odd points, one could examine for each grid point which neighboring points are most appropriate for prediction. At this moment, this reorganization idea soon gets stuck into practical troubles, but it certainly merits further investigation.

7.3.4 Starting from a first-generation solution

We know that if the transform neglects the grid structure, the result reflects the irregularity of the grid. The result is non-smooth, which means that it has no sparse representation in a second-generation basis. Apart from that, the result is fairly reliable, in the sense that bias is restricted by the Riesz-constants of the transform: if we are thresholding in the wavelet domain, we know what we are doing in the original signal domain. Let $w^{(1)}$ be the second generation wavelet coefficients of this first generation solution $y^{(1)}$. Our objective is to find a sparsely represented signal close to $y^{(1)}$. To this end, we use the thresholded coefficients w_λ of the second-generation transform of the noise.

If a coefficient w_i corresponds to a wavelet that lies on an interval where the second-generation solution y_λ shows no bias, we can choose as output:

$$\hat{w}_i = w_{\lambda i}$$

To do so, we have to define in which data points y_λ is biased and we have to mark the coefficients that correspond to these points. We say that $y_{\lambda i}$ is biased if

$$|y_{\lambda i} - y_i^{(1)}| > \hat{\sigma},$$

where:

$$\hat{\sigma} = \sqrt{\frac{1}{N} \sum_{i=1}^{N} (y_i - y_i^{(1)})^2}$$

is an estimate of the noise variance (we suppose that the noise is stationary). This definition is subject to the remaining noise and the irregular grid effects in $y_i^{(1)}$. Because we expect that bias has typically a range of more than one data point, we first filter out isolated points that were classified as biased, before the actual marking of the corresponding wavelet coefficients.

For all these marked coefficients w_i we compute the value of

$$(w_{\lambda i} - w_i^{(1)})^2 \int_{-\infty}^{\infty} \psi_i^2(x) \, dx = (w_{\lambda i} - w_i^{(1)})^2 (W^T T_J W)_{ii},$$

which quantifies the effect on the output if we replace $w_i^{(1)}$ by $w_{\lambda i}$. If we compute the sum of these effects over all marked coefficients, we see that a few of them are responsible for the major part of the bias. These coefficients, together with the untouched scaling coefficients, keep their value $w_i^{(1)}$. All others undergo the same procedure as the unmarked coefficients.

This procedure eliminates large noise coefficients that do not interfere with biased reconstruction points. This is how the algorithm gets rid of most hidden noise components.

Instead of marking wavelet coefficients that correspond to intervals with bias, we can also compute for all coefficients the value of:

$$B_i = (w_{\lambda i} - w_i^{(1)})^2 \int_{-\infty}^{\infty} \psi_i^2(x) \chi_{\text{bias}}(x) \, dx.$$

$\chi_{\text{bias}}(x)$ is an indicator function which is one on all intervals with bias. The above value measures the participation of w_i in the bias. If M is a diagonal matrix with $M_{kk} = 1$ if the corresponding data point x_k has been marked as biased and $M_{kk} = 0$ otherwise, B_i can be computed as:

$$B_i = (w_{\lambda i} - w_i^{(1)})^2 (W^T (T_J M) W)_{ii}.$$

Marking the coefficients with the highest values gives results very close to the first selecting procedure.

7.3.5 The proposed algorithm

The objective of the algorithm is to combine the smooth reconstruction of a second-generation procedure with the reliable estimation of the classical transform. We call $\tilde{W}$ and W the forward and inverse second generation transform, as before, and $\tilde{U}$ and U are the transform matrices if we do not take into account the grid structure. The algorithm goes as follows:

1. Compute $w = \tilde{W}y$ and $u = \tilde{U}y$.
2. Compute the structure of the covariance matrix of the wavelet coefficients:

$$S = \tilde{W}Q\tilde{W}^T,$$

   and similarly for $\tilde{U}$. Q contains the covariance matrix of the input (up to constant; we do not use an estimate of the noise variance). We assume that the noise is stationary and uncorrelated: $Q = I$. In that case, the computation of S has linear complexity.
3. Normalize the coefficients with these variances and select for both sets of wavelet coefficients a threshold λ and μ, e.g. by minimizing GCV$_w(\lambda)$ and GCV$_u(\mu)$. And apply a soft-threshold to get the thresholded vectors w_λ and u_μ.
4. Compute $y^{(1)} = Uu_\mu$ and $w^{(1)} = \tilde{W}y^{(1)}$. Use the not further transformed scaling coefficients in $w^{(1)}$ as an estimate for the corresponding noise-free scaling coefficients. Replace the noisy scaling coefficients in w_λ by these values and compute $y_\lambda = Ww_\lambda$.
5. Estimate the noise standard deviation by:

$$\hat{\sigma} = \sqrt{\frac{1}{N}\sum_{i=1}^{N}(y_i - y_i^{(1)})^2}.$$

   Among all data points $i = 1, \ldots, N$ we mark those for which $|y_{\lambda i} - y_i^{(1)}| > \hat{\sigma}$, $y_{\lambda i}$ as biased. We filter out isolated labels, since we consider bias as a more-than-one-point phenomenon. We mark all coefficients corresponding to basis functions on intervals with bias.
6. Compute the 2-norm of each basis function, these can be found in the diagonal of the matrix $W^T T_J W$. The computation of this diagonal is of linear complexity.
7. Among all marked coefficients $w_{\lambda i}$, unmark those for which

$$(w_{\lambda i} - w_i^{(1)})^2 (W^T T_J W)_{ii}$$

   is too small. Make sure that all scaling coefficients are marked.
8. For all coefficients $i = 1, \ldots, N$ select the appropriate value:
   a) If a coefficient is marked, let: $\hat{w}_i = w_i^{(1)}$,
   b) for the others, select: $\hat{w}_i = w_{\lambda i}$.
9. The output is:

$$\hat{y} = W\hat{w}.$$

This algorithm requires 3 forward and 3 inverse transforms, but the order of complexity is still linear. The computation of $W^T T_J W$ and $\tilde{W}Q\tilde{W}^T$ are the most time consuming steps.

7.3.6 Results and discussion

Figure 7.8 contains a plot of the result of the proposed algorithm. It is smooth and close to the noise-free signal. Figure 7.9 focusses on a detail and illustrates the importance of the grid: if neglect the grid structure, the result is non-smooth. The wavelet transform used a cubic interpolation as prediction filter, followed by a two taps update-filter, designed to create the dual wavelets with two vanishing moments. The input signal had $N = 2048$ data points, and we leave 8 scaling coefficients untransformed, and so untouched by the threshold. For the reconstruction, only 18 from the 2048 wavelet coefficients, including the 8 scaling coefficients, were taken from $w^{(1)}$, all the others were based on the thresholded second generation coefficients.

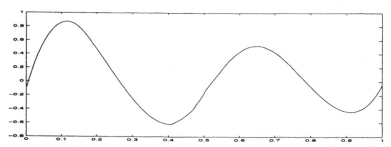

Figure 7.8. Result of the proposed algorithm. It is smooth and close to the noise-free signal.

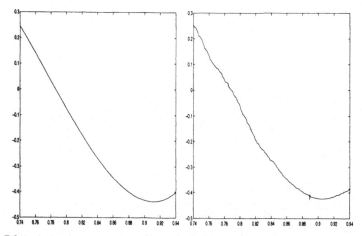

Figure 7.9. Left: detail of Figure 7.8. Right: the reconstruction of classical procedure (no grid structure) on the same interval. This reconstruction carries the irregularity of the grid.

7.3.6 Results and discussion

Figure 7.8 contains a plot of the result of the proposed algorithm. It is smooth and close to the noise-free signal. Figure 7.9 focuses on a detail and illustrates the importance of the grid. If neglect the grid structure the result is non-smooth. The wavelet transform uses a cubic interpolation as prediction filter, followed by a two taps update filter designed to create the dual wavelet is with two vanishing moments. The input signal had $N = 2048$ data points, and we leave 8 scaling coefficients un-smoothed, all other computed by the threshold. For the reconstruction only 18 from the 2048 used interpolating prediction coefficients, were taken.

Figure 7.8. Left: Proposed algorithm result with grid. Dots are the noisy signal.

Figure 7.9. Left: detail of Figure 7.8. Right: the reconstruction of classical procedure (no grid structure) on the same interval. This reconstruction carries the irregularity of the grid.

Bibliography

[1] F. Abramovich, T. C. Bailey, and Th. Sapatinas. Wavelet analysis and its statistical applications. *The Statistician - Journal of the Royal Statistical Society, Ser. D*, To appear, 2000.

[2] F. Abramovich and Y. Benjamini. Adaptive thresholding of wavelet coefficients. *Computational Statistics and Data Analysis*, 22:351–361, 1996.

[3] F. Abramovich, Y. Benjamini, D. L. Donoho, and I. M. Johnstone. Adapting to unknown sparsity by controlling the false discovery rate. Technical Report 2000-19, Department of Statistics, Stanford University, 2000.

[4] F. Abramovich, F. Sapatinas, and B. W. Silverman. Wavelet thresholding via a Bayesian approach. *Journal of the Royal Statistical Society, Series B*, 60:725–749, 1998.

[5] F. Abramovich and B. W. Silverman. Wavelet decomposition approaches to statistical inverse problems. *Biometrika*, 85:115–129, 1998.

[6] A. N. Akansu and R. A. Haddad. *Multiresolution Signal Decomposition: Transforms, Subbands, and Wavelets*. Academic Press, 1250 Sixth Ave., San Diego, CA 92101-4311, 1992.

[7] N. S. Altman. Kernel smoothing of data with correlated errors. *J. Amer. Statist. Assoc.*, 85:749–759, Sep 1990.

[8] U. Amato and D. T. Vuza. Besov regularization, thresholding and wavelets for smoothing data. *Numer. Funct. Anal. Optimization*, 18:461–493, 1997.

[9] U. Amato and D. T. Vuza. Wavelet approximation of a function from samples affected by noise. *Revue Roum. Math. Pures Appl.*, 42(7–8):481–493, 1997.

[10] F. Anscombe. The transformation of Poisson, binomial and negative binomial data. *Biometrika*, 35:246–254, 1948.

[11] J.-P. Antoine. The continuous wavelet transform in image processing. *CWI Q.*, 11(4):323–345, 1998.

[12] J.-P. Antoine, P. Carrette, R. Murenzi, and B. Piette. Image analysis with two-dimensional continuous wavelet transform. *Signal Processing*, 31(3):241–272, 1993.

[13] A. Antoniadis and J. Fan. Regularized wavelet approximations. *J. Amer. Statist. Assoc.*, 1999. Submitted.

[14] A. Antoniadis and I. Gijbels. Detecting abrupt changes by wavelet methods. *J. Nonparam. Statistics*, to appear, 2000.

[15] A. Antoniadis, G. Grégoire, and W McKeague. Wavelet methods for curve estimation. *J. Amer. Statist. Assoc.*, 89:1340–1353, 1994.

[16] A. Antoniadis, G. Grégoire, and P. Vial. Random design wavelet curve smoothing. *Statistics and Probability Letters*, 35:225–232, 1997.

[17] A. Antoniadis and D.-T. Pham. Wavelet regression for random or irregular design. *Computational Statistics and data analysis*, 28(4):333–369, 1998.

[18] M. Antonini, M. Barlaud, P. Mathieu, and I. Daubechies. Image coding using the wavelet transform. *IEEE Transactions on Image Processing*, 1(2):205–220, 1992.

[19] A. Arneodo, E. Bacry, S. Jaffard, and J.F. Muzy. Singularity spectrum of multifractal functions involving oscillating singularities. *J. Fourier Anal. Appl.*, 4(2):159–174, 1998.

[20] R. Baraniuk. Optimal tree approximation with wavelets. In M. A. Unser, A. Aldroubi, and Laine A. F., editors, *Wavelet Applications in Signal and Image Processing VII*, volume 3813 of *SPIE Proceedings*, pages 196–207, July 1999.

[21] O. E. Barndorff-Nielsen and D. R. Cox. *Asymptotic Techniques for Use in Statistics*. Chapman and Hall, 11 New Fetter Lane, London EC4P 4EE, U.K., 1989.

[22] M. G. Bello. A combined Markov Random Field and wave-packet transform-based approach for image segmentation. *IEEE Transactions on Image Processing*, 3(6):834–846, 1994.

[23] Y. Benjamini and Y. Hochberg. Controlling the false discovery rate: A practical and powerful approach to multiple testing. *Journal of the Royal Statistical Society, Series B*, 57:289–300, 1995.

[24] K. Berkner and R. O. Wells, Jr. Correlation-dependent model for denoising via nonorthogonal wavelet transforms. Technical report, C. M. L., Dept. Mathematics, Rice University, June 1998.

[25] J. E. Besag. Spatial interaction and the spatial analysis of lattice systems. *Journal of the Royal Statistical Society, Series B*, 36:192–236, 1974.

[26] Ch. Blatter. *Wavelets. A primer*. Natick, MA: A K Peters, 1998.

[27] C. A. Bouman and M. Shapiro. A multiscale random field model for Bayesian image segmentation. *IEEE Transactions on Image Processing*, 3(2):162–177, 1994.

[28] L. Breiman, J. H. Friedman, R. A. Olshen, and C. J. Stone. *Classification and regression trees (CART)*. Wadsworth, Monterey, CA, USA, 1984.

[29] S. C. Brenner and L. R. Scott. *The Mathematical Theory of Finite Element Methods*. Springer-Verlag, 175 Fifth Avenue, New York 10010, USA, 1994.

[30] A.G. Bruce and H.-Y. Gao. Understanding WaveShrink: Variance and bias estimation. *Biometrika*, 83:727–745, 1996.

[31] A.G. Bruce and H.-Y. Gao. Waveshrink with firm shrinkage. *Statistica Sinica*, 4:855–874, 1997.

[32] C. S. Burrus, R. A. Gopinath, and H. Guo. *Introduction to Wavelets and Wavelet Transforms.* Prentice Hall, Upper Saddle River, New Jersey 07458, 1998.

[33] T. Cai and L.D. Brown. Wavelet shrinkage for nonequispaced samples. *Annals of Statistics*, 26(5):1783–1799, 1998.

[34] T. T. Cai. Adapative wavelet estimation: a block thresholding and oracle inequality approach. *Annals of Statistics*, 27:898–924, 1999.

[35] T. T. Cai and B. W. Silverman. Incorporating information on neighboring coefficients into wavelet estimation. Technical Report 98-13, Department of Statistics, Purdue University, 1998.

[36] R. Calderbank, I. Daubechies, W. Sweldens, and B.-L. Yeo. Wavelet transforms that map integers to integers. *Appl. Comput. Harmon. Anal.*, 5(3):332–369, 1998.

[37] E. Candes. *Ridgelets: theory and applications.* PhD thesis, Department of Statistics, Stanford University, August 1998.

[38] E. J. Candes and D. L. Donoho. Ridgelets: a key to higher dimensional intermittency? *Journal of the Royal Statistical Society, Series A*, 357:2495–2509, 1999.

[39] E. J. Candes and D. L. Donoho. Curvelets - a surprisingly effective nonadaptive representation for objects with edges. Technical report, Department of Statistics, Stanford University, 2000.

[40] A. Chambolle, R. A. DeVore, N.-Y. Lee, and B. J. Lucier. Nonlinear wavelet image processing: Variational problems, compression, and noise removal through wavelet shrinkage. *IEEE Transactions on Image Processing*, 7(3):319–355, March 1998.

[41] G. Chang, B. Yu, and M. Vetterli. Adaptive wavelet thresholding for image denoising and compression. *submitted*, 1999.

[42] G. Chang, B. Yu, and M. Vetterli. Spatially adaptive wavelet thresholding based on context modeling for image denoising. *submitted*, 1999.

[43] G. Chang, B. Yu, and M. Vetterli. Wavelet thresholding for multiple noisy image copies. *IEEE Transactions on Image Processing*, To appear, 1999.

[44] P. Charbonnier, L. Blanc-Féraud, and M. Barlaud. Noisy image restoration using multiresolution Markov Random Fields. *Journal of Visual Communication and Image Representation*, 3(4):338–346, 1992.

[45] S. Chen and D. L. Donoho. Atomic decomposition by basis pursuit. Technical Report 479, Department of Statistics, Stanford University, May 1995.

[46] H. Chipman, E. Kolaczyk, and R. McCulloch. Adaptive Bayesian wavelet shrinkage. *J. Amer. Statist. Assoc.*, 92:1413–1421, 1997.

[47] H. Choi and R. G. Baraniuk. Wavelet statistical models and besov spaces. In M. A. Unser, A. Aldroubi, and A. F. Laine, editors, *Wavelet Applications in Signal and Image Processing VII*, volume 3813 of *SPIE Proceedings*, pages 489–501, July 1999.

[48] M. Clyde, G. Parmigiani, and B. Vidakovic. Multiple shrinkage and subset selection in wavelets. *Biometrika*, 85:391–401, 1998.

[49] A. Cohen, I. Daubechies, and J. Feauveau. Bi-orthogonal bases of compactly supported wavelets. *Comm. Pure Appl. Math.*, 45:485–560, 1992.

[50] I. Cohen, I. Raz, and D. Malah. Orthonormal shift-invariant adaptive local trigonometric decomposition. *Signal Processing*, 57(1):43–64, 1997.

[51] I. Cohen, I. Raz, and D. Malah. Orthonormal shift-invariant wavelet packet decomposition and representation. *Signal Processing*, 57(3):251–270, 1997.

[52] R. Coifman and Y. Meyer. Remarques sur l'analyse de Fourier à fenêtre. *C. R. Acad. Sci. Paris Sér. I Math.*, I(312):259–261, 1991.

[53] R. R. Coifman and D. L. Donoho. Translation-invariant de-noising. In A. Antoniadis and G. Oppenheim, editors, *Wavelets and Statistics*, Lecture Notes in Statistics, pages 125–150, 1995.

[54] R. R. Coifman, Y. Meyer, and V. Wickerhauser. Size properties of wavelet packets. In M. B. Ruskai, G. Beylkin, R. Coifman, I. Daubechies, S. Mallat, Y. Meyer, and L. Raphael, editors, *Wavelets and their Applications*, pages 453–470. Jones and Bartlett, Boston, 1992.

[55] R. R. Coifman and M. L. Wickerhauser. Entropy based algorithms for best basis selection. *IEEE Transactions on Information Theory*, 38(2):713–718, 1992.

[56] P. Craven and G. Wahba. Smoothing noisy data with spline functions. *Numerische Mathematik*, 31:377–403, 1979.

[57] M. S. Crouse, R.D. Nowak, and R. G. Baraniuk. Wavelet-based signal processing using hidden markov models. *IEEE Transactions on Signal Processing*, 46, Special Issue on Wavelets and Filterbanks:886–902, 1998.

[58] W. Dahmen, A. J. Kurdila, and P. Oswald. *Multiscale wavelet methods for partial differential equations*. Academic press New York, 1997.

[59] I. Daubechies. *Ten Lectures on Wavelets*. CBMS-NSF Regional Conf. Series in Appl. Math., Vol. 61. Society for Industrial and Applied Mathematics, Philadelphia, PA, 1992.

[60] I. Daubechies, I. Guskov, P. Schröder, and W. Sweldens. Wavelets on irregular point sets. *Phil. Trans. R. Soc. Lond. A*, To be published.

[61] I. Daubechies, I. Guskov, and W. Sweldens. Regularity of irregular subdivision. *Constructive Approximation*, 15:381–426, 1999.

[62] I. Daubechies and W. Sweldens. Factoring wavelet transforms into lifting steps. *J. Fourier Anal. Appl.*, 4(3):245–267, 1998.

[63] P. De Gersem, B. De Moor, and M Moonen. Applications of the continuous wavelet transform in the processing of musical signals. In *Proc. of the 13th International Conference on Digital Signal Processing (DSP97)*, pages 563–566, 1997.

[64] V. Delouille. Nonparametric regression estimation using design-adapted wavelets. Master's thesis, Institut de Statistique, UCL, Belgium, 1999.

[65] B. Delyon and A. Juditsky. On the computation of wavelet coefficients. *J. of Approx. Theory*, 88:47–79, 1997.

[66] A. P. Dempster, N. M. Laird, and D. B. Rubin. Maximum likelihood from incomplete data via the em algorithm. *Journal of the Royal Statistical Society, Series B*, 39:1–38, 1977.

[67] R. A. DeVore, B. B. Jawerth, and V. Popov. Compression of wavelet decompositions. *Amer. J. Math.*, 114:737–785, 1992.

[68] R. A. DeVore, B. Jawerth, and B. J. Lucier. Image compression through wavelet transform coding. *IEEE Transactions on Information Theory*, 38(2):719–746, 1992.

[69] R. A. DeVore and V Popov. Interpolation of Besov spaces. *Trans. Amer. Math. Soc.*, 305:397–414, 1988.

[70] D. L. Donoho. Wavelet shrinkage and W.V.D. – a ten-minute tour. In Y. Meyer and S. Roques, editors, *Progress in Wavelet Analysis and Applications*, pages 109–128. Editions Frontières: Gif-sur-Yvette, 1993.

[71] D. L. Donoho. De-noising by soft-thresholding. *IEEE Transactions on Information Theory*, 41(3):613–627, May 1995.

[72] D. L. Donoho. Nonlinear solution of linear inverse problems by wavelet-vaguelette decomposition. *Appl. Comput. Harmon. Anal.*, 2(2):101–126, 1995.

[73] D. L. Donoho. CART and best-ortho-basis: a connection. *Annals of Statistics*, 25(5):1870–1911, 1997.

[74] D. L. Donoho. Orthonormal ridgelets and linear singularities. *SIAM J. Math. Anal.*, 31:1062–1099, 2000.

[75] D. L. Donoho and I. M. Johnstone. Ideal denoising in an orthogonal basis chosen from a library of bases. *C. R. Acad. Sci. I - Math.*, 319:1317–1322, 1994.

[76] D. L. Donoho and I. M. Johnstone. Ideal spatial adaptation via wavelet shrinkage. *Biometrika*, 81:425–455, 1994.

[77] D. L. Donoho and I. M. Johnstone. Minimax risk over ℓ_p-balls for ℓ_q-error. *Prob. Theory Relat. Fields*, 99(2):277–303, 1994.

[78] D. L. Donoho and I. M. Johnstone. Adapting to unknown smoothness via wavelet shrinkage. *J. Amer. Statist. Assoc.*, 90:1200–1224, 1995.

[79] D. L. Donoho and I. M. Johnstone. Neo-classical minimax theorems, thresholding, and adaptive function estimation. *Bernoulli*, 2(1):39–62, 1996.

[80] D. L. Donoho and I. M. Johnstone. Minimax estimation via wavelet shrinkage. *Annals of Statistics*, 26:879–921, 1998.

[81] D. L. Donoho, I. M. Johnstone, G. Kerkyacharian, and D. Picard. Wavelet shrinkage: Asymptopia? *Journal of the Royal Statistical Society, Series B*, 57(2):301–369, 1995.

[82] R. Fletcher. *Practical methods of optimization*. Wiley-interscience publications. Wiley, Chichester, 1981.

[83] M. W. Frazier. *An Introduction to Wavelets through Linear Algebra*. Springer-Verlag, New York, 1999.

[84] H.-Y. Gao. Wavelet shrinkage denoising using the non-negative garrote. *Journal of Computational and Graphical Statistics*, 7(4):469–488, December 1998.

[85] S. Geman and D. Geman. Stochastic relaxation, Gibbs distributions, and the Bayesian restoration of images. *IEEE Transactions on Pattern Analysis and Machine Intelligence*, 6(6):721–741, 1984.

[86] J. W. Gibbs. *Elementary principles in statistical mechanics*. Yale University Press, New Haven, 1902. Reprinted by Ox Bow Press, Woodbridge, 1981.

[87] M. Girardi and W. Sweldens. A new class of unbalanced Haar wavelets that form an unconditional basis for L_p on general measure spaces. *J. Fourier Anal. Appl.*, 3(4):457–474, 1997.

[88] S. Goedecker. *Wavelets and their application for the solution of partial differential equations in physics*. Cahiers de Physique. 4. Presses Politechniques et Universitaires Romandes, Lausanne, 1998.

[89] S. Goedecker and O. Ivanov. Solution of multiscale partial differential equations using wavelets. *Computers in Physics*, 12(6):548–555, 1998.

[90] A. Haar. Zur Theorie der orthogonalen Funktionen-Systeme. *Math. Ann.*, 69:331–371, 1910.

[91] P. Hall, G. Kerkyarcharian, and D. Picard. Block threshold rules for curve estimation using kernel and wavelet methods. *Annals of Statistics*, 26:922–942, 1998.

[92] P. Hall and I. Koch. On the feasibility of cross-validation in image analysis. *SIAM J. Appl. Math.*, 52(1):292–313, 1992.

[93] P. Hall and P. Patil. Formulae for mean integrated squared error of nonlinear wavelet-based density estimators. *Annals of Statistics*, 23(3):905–928, 1995.

[94] P. Hall and P. Patil. Effect of threshold rules on performance of wavelet-based curve estimators. *Stat. Sin.*, 6(2):331–345, 1996.

[95] P. Hall and P. Patil. On the choice of smoothing parameter, threshold and truncation in nonparametric regression by non-linear wavelet methods. *Journal of the Royal Statistical Society, Series B*, 58(2):361–377, 1996.

[96] P. Hall and B. A. Turlach. Interpolation methods for nonlinear wavelet regression with irregularly spaced design. *Annals of Statistics*, 25(5):1912–1925, 1997.

[97] M. Hansen and B. Yu. Wavelet thresholding via MDL: simultaneous denoising and compression. *submitted*, 1999.

[98] J. D. Hart. Kernel regression estimation with time series errors. *Journal of the Royal Statistical Society, Series B*, 53(1):173–187, 1991.

[99] J. D. Hart. Automated kernel smoothing of dependent data by using time series cross-validation. *Journal of the Royal Statistical Society, Series B*, 56(3):529–542, 1994.

[100] J. Heikkinen and H. Högmander. Fully Bayesian approach to image restoration with an application in biogeography. *Applied Statistics*, 43(4):569–582, 1994.

[101] H. M. Hudson. A natural identity for exponential families with applications in multiparameter estimation. *Annals of Statistics*, 6(3):473–484, 1978.

[102] E. Ising. Beitrag zur Theorie des Ferromagnetismus. *Zeitschrift für Physik*, 31:253–258, 1925.

[103] S. Jaffard. Pointwise smoothness, two-microlocalisation and wavelet coefficients. *Publicacions Matemàtiques*, 35:155–168, 1991.

[104] M. Jansen and A. Bultheel. Geometrical priors for noisefree wavelet coefficient configurations in image de-noising. In P. Müller and B. Vidakovic, editors, *Bayesian inference in wavelet based models*, pages 223–242. Springer-Verlag, 1999.

[105] M. Jansen and A. Bultheel. Multiple wavelet threshold estimation by generalized cross validation for images with correlated noise. *IEEE Transactions on Image Processing*, 8(7):947–953, July 1999.

[106] M. Jansen, M. Malfait, and A. Bultheel. Generalized cross validation for wavelet thresholding. *Signal Processing*, 56(1):33–44, January 1997.

[107] B. Jawerth and W. Sweldens. An overview of wavelet based multiresolution analyses. *SIAM Review*, 36(3):377–412, 1994.

[108] I. M. Johnstone. Minimax bayes, asymptotic minimax and sparse wavelet priors. In S. Gupta and J. Berger, editors, *Statistical Decision Theory and Related Topics, V*, pages 303–326. Springer-Verlag, 1994.

[109] I. M. Johnstone and B. W. Silverman. Wavelet threshold estimators for data with correlated noise. *Journal of the Royal Statistical Society, Series B*, 59:319–351, 1997.

[110] G. Kaiser. *A Friendly Guide to Wavelets*. Birkhäuser, 675 Massachusetts Ave., Cambridge, MA 02139, U.S.A., 1994.

[111] E. D. Kolaczyk. A wavelet shrinkage approach to tomographic image reconstruction. *J. Amer. Statist. Assoc.*, 91:1079–1090, 1996.

[112] E. D. Kolaczyk. Wavelet shrinkage estimation of certain Poisson intensity signals using corrected thresholds. *Stat. Sin.*, 9(1):119–135, 1999.

[113] A. Kovac and B. W. Silverman. Extending the scope of wavelet regression methods by coefficient-dependent thresholding. *J. Amer. Statist. Assoc.*, 95:172–183, 2000.

[114] J. Kovačević and W. Sweldens. Wavelet families of increasing order in arbitrary dimensions. *IEEE Transactions on Image Processing*, 1999.

[115] J. Kovačević and M. Vetterli. Nonseparable multidimensional perfect reconstruction filter banks and wavelet bases for $\mathbf{R}^n$. *IEEE Transactions on Information Theory*, 38(2):533–555, March 1992.

[116] F. Labaere, P. Vuylsteke, P. Wambacq, E. Schoeters, and C. Fivez. Primitive-based contrast enhancement method. In M. Loew and K. Hanson, editors, *Medical Imaging 1996: Image Processing*, volume 2710 of *SPIE Proceedings*, pages 811–820, April 1996.

[117] G. Lang, H. Guo, J. E. Odegard, C. S. Burrus, and R. O. Wells. Noise reduction using an undecimated discrete wavelet transform. *IEEE Signal Processing Letters*, 3(1):10–12, 1996.

184   Bibliography

[118] M. Lang, H. Guo, J. E. Odegard, C. S. Burrus, and R. O. Wells. Nonlinear processing of a shift-invariant discrete wavelet transform (dwt) for noise reduction. In H. H. Szu, editor, *Wavelet Applications II*, pages 640–651, April 1995.

[119] M. R. Leadbetter, G. Lindgren, and H. Rootzén. *Extremes and Related Properties of Random Sequences and Processes*. Springer Series in Statistics. Springer, 175 Fifth Avenue, New York 10010, USA, 1983.

[120] A. K. Louis, P. Maaß, and A. Rieder. *Wavelets: Theory and Applicaltions*. John Wiley & Sons, 605 Third Avenue, New York, NY 10158-0012, USA, 1997.

[121] J. Lu, D. M. Healy Jr., and J. B. Weaver. Contrast enhancement of medical images using multiscale edge representation. *Optical Engineering*, special issue on Adaptive Wavelet Tansforms:pages 1251–1261, July 1994.

[122] M. R. Luettgen, W. C. Karl, A. S. Willsky, and R. R. Tenney. Multiscale representations of Markov Random Fields. *IEEE Transactions on Signal Processing*, 41(12):3377–3395, December 1993.

[123] M. Malfait. *Stochastic Sampling and Wavelets for Bayesian Image Analysis*. PhD thesis, Department of Computer Science, K.U.Leuven, Belgium, 1995.

[124] M. Malfait. Using wavelets to suppress noise in biomedical images. In A. Aldroubi and M. Unser, editors, *Wavelets in Medicine and Biology*, Chapter 8, pages 191–208. CRC Press, 1995.

[125] M. Malfait and D. Roose. Wavelet based image denoising using a markov random field a priori model. *IEEE Transactions on Image Processing*, 6(4):549–565, 1997.

[126] S. Mallat and W. L. Hwang. Singularity detection and processing with wavelets. *IEEE Transactions on Information Theory*, 38(2):617–643, 1992.

[127] S. Mallat and Z. Zhang. Matching pursuits with time-freuency dictionaries. *IEEE Transactions on Signal Processing*, 41(12):3397–3415, 1993.

[128] S. Mallat and S. Zhong. Characterization of signals from multiscale edges. *IEEE Transactions on Pattern Analysis and Machine Intelligence*, 14:710–732, 1992.

[129] S. G. Mallat. A theory for multiresolution signal decomposition: The wavelet representation. *IEEE Transactions on Pattern Analysis and Machine Intelligence*, 11(7):674–693, 1989.

[130] S. G. Mallat. *A Wavelet Tour of Signal Processing*. Academic Press, 525 B Street, Suite 1900, San Diego, CA, 92101-4495, USA, 1998.

[131] H. S. Malvar. Lapped transforms for efficient transform/subband coding. *IEEE Trans. Acoust. Speech Signal Process.*, 38:969–978, 1990.

[132] H. S. Malvar and D. H. Staelin. The LOT: Transform coding without blocking effects. *IEEE Trans. Acoust. Speech Signal Process.*, 37:553–559, 1989.

[133] J. S. Marron, S. Adak, I. M. Johnstone, M. H. Neumann, and P. Patil. Exact risk analysis of wavelet regression. *Journal of Graphical and Computational Statistics*, 7:278–309, 1998.

[134] N. Metropolis, A. Rosenbluth, M. Rosenbluth, A. Teller, and E. Teller. Equation of state calculations by fast computing machines. *Journal of Chemical Physics*, 21:1087–1092, 1953.

[135] G. P. Nason. Wavelet regression by cross validation. Preprint, Department of Mathematics, University of Bristol, UK, 1994.

[136] G. P. Nason. Wavelet shrinkage using cross validation. *Journal of the Royal Statistical Society, Series B*, 58:463–479, 1996.

[137] G. P. Nason and B. W. Silverman. The stationary wavelet transform and some statistical applications. In A. Antoniadis and G. Oppenheim, editors, *Wavelets and Statistics*, Lecture Notes in Statistics, pages 281–299, 1995.

[138] G. P. Nason and R. von Sachs. Wavelets in time series analysis. *Philosophical Transactions of the Royal Society London, Series A: Mathematical, Physical and Engineering Sciences*, 357:2511–2526, 1999.

[139] G. P. Nason, R. von Sachs, and G. Kroisandt. Wavelet processes and adaptive estimation of the evolutionary wavelet spectrum. *Journal of the Royal Statistical Society, Series B*, 62:271–292, 2000.

[140] M. H. Neumann and R. von Sachs. Wavelet thresholding in anisotropic function classes and application to adaptive estimation of evolutionary spectra. *Annals of Statistics*, 25:38–76, 1997.

[141] Y. Nievergelt. *Wavelets made easy*. Birkhaeuser, Boston, MA, 1999.

[142] R.D. Nowak and R. G. Baraniuk. Optimal weighted highpass filters using multiscale analysis. *IEEE Transactions on Image Processing*, 1996. submitted.

[143] R.D. Nowak and R. G. Baraniuk. Wavelet-domain filtering for photon imaging systems. *IEEE Transactions on Image Processing*, 1998. submitted.

[144] R. T. Ogden. *Essential Wavelets for Statistical Applications and Data Analysis*. Birkhauser, Boston, 1997.

[145] T. Ogden and E. Parzen. Change-point approach to data analytic wavelet thresholding. *Statistics and Computing*, 6:93–99, 1996.

[146] T. Ogden and E. Parzen. Data dependent wavelet thresholding in nonparametric regression with change-point applications. *Computational Statistics and Data Analysis*, 22(1):53–70, 1996.

[147] M. Pensky and B. Vidakovic. On non-equally spaced wavelet regression. Preprint, Duke University, Durham, NC, 1998.

[148] J.-C. Pesquet, H. Krim, and H. Carfantan. Time invariant orthonormal wavelet representations. *IEEE Transactions on Signal Processing*, 44(8):1964–1970, 1996.

[149] L. Prasad and S. S. Iyengar. *Wavelet Analysis with Applications to Image Processing*. CRC Press, 2000 Corporate Blvd., N.W., Boca Raton, Florida 33431, 1997.

[150] M. Raimondo. Minimax estimation of sharp change points. *Annals of Statistics*, 26:1379–1397, 1998.

[151] F. Ruggeri and B. Vidakovic. A Bayesian decision theoretic approach to the choice of thresholding parameter. *Statistica Sinica*, 9:183–197, 1999.

[152] S. Sardy, D.B. Percival, A.G. Bruce, H-Y. Gao, and W. Stuetzle. Wavelet de-noising for unequally spaced data. *Statistics and Computing*, 9:65–75, 1999.

[153] J. Simoens and S. Vandewalle. On the stability of wavelet bases in the lifting scheme. TW Report 306, Department of Computer Science, Katholieke Universiteit Leuven, Belgium, 2000.

[154] E. Simoncelli. Modeling the joint statistics of images in the wavelet domain. In M. A. Unser, A. Aldroubi, and Laine A. F., editors, *Wavelet Applications in Signal and Image Processing VII*, volume 3813 of *SPIE Proceedings*, pages 206–214, July 1999.

[155] E. P. Simoncelli and E.H. Adelson. Noise removal via Bayesian wavelet coring. In *proceedings 3rd International Conference on Image Processing*, September 1996.

[156] E.P. Simoncelli and E.H. Adelson. Non-separable extensions of quadrature mirror filters to multiple dimensions. *Proceedings of the IEEE*, 78:652–664, April 1990.

[157] C. Stein. Estimation of the mean of a multivariate normal distribution. *Annals of Statistics*, 9(6):1135–1151, 1981.

[158] G. Strang and T. Nguyen. *Wavelets and filter banks*. Wellesley-Cambridge Press, Box 812060, Wellesley MA 02181, fax 617-253-4358, 1996.

[159] W. Sweldens. *The Construction and Application of Wavelets in Numerical Analysis*. PhD thesis, Department of Computer Science, K.U.Leuven, Belgium, 1994.

[160] W. Sweldens. The lifting scheme: A new philosophy in biorthogonal wavelet constructions. In A. F. Laine and M. Unser, editors, *Wavelet Applications in Signal and Image Processing III*, pages 68–79. Proc. SPIE 2569, 1995.

[161] W. Sweldens. The lifting scheme: A custom-design construction of biorthogonal wavelets. *Appl. Comput. Harmon. Anal.*, 3(2):186–200, 1996.

[162] W. Sweldens. The lifting scheme: A construction of second generation wavelets. *SIAM J. Math. Anal.*, 29(2):511–546, 1997.

[163] W. Sweldens and R. Piessens. Quadrature formulae and asymptotic error expansions for wavelet approximations of smooth functions. *SIAM J. Numer. Anal.*, 31(4):1240–1264, 1994.

[164] W. Sweldens and P. Schröder. Building your own wavelets at home. In *Wavelets in Computer Graphics*, ACM SIGGRAPH Course Notes, pages 15–87. ACM, 1996.

[165] A. Teolis. *Computational signal processing with wavelets*. Applied and Numerical Harmonic Analysis. Birkhaeuser, Boston, MA, 1998.

[166] G. Uytterhoeven. *Wavelets: software and applications*. PhD thesis, Department of Computer Science, K.U.Leuven, Belgium, April 1999.

[167] G. Uytterhoeven and A. Bultheel. The Red-Black wavelet transform. TW Report 271, Department of Computer Science, Katholieke Universiteit Leuven, Belgium, December 1997.

[168] G. Uytterhoeven, F. Van Wulpen, M. Jansen, D. Roose, and A. Bultheel. WAILI: Wavelets with Integer Lifting. TW Report 262, Department of Computer Science, Katholieke Universiteit Leuven, Belgium, July 1997.

[169] D. Vandermeulen. *Methods for registration, interpolation and interpretation of three-dimensional medical image data for use in 3-D display, 3-D modelling and therapy planning.* PhD thesis, K.U.Leuven, 1991.

[170] M. Vetterli and C. Herley. Wavelets and filter banks: theory and design. *IEEE Transactions on Signal Processing*, 40(9):2207–2232, 1992.

[171] B. Vidakovic. Nonlinear wavelet shrinkage with Bayes rules and Bayes factors. *J. Amer. Statist. Assoc.*, 93:173–179, 1998.

[172] B. Vidakovic. *Statistical Modeling by Wavelets.* Wiley Series in Probability and Mathematical Statistics - Applied Probability and Statistics Section. John Wiley & Sons, 605 Third Avenue, New York, NY 10158-0012, USA, 1999.

[173] R. von Sachs and B. MacGibbon. Non-parametric curve estimation by wavelet thresholding with locally stationary errors. *Scandinavian Journal of Statistics*, 27:475–499, 2000.

[174] G. Wahba. *Spline Models for Observational Data*, chapter 4, pages 45–65. CBMS-NSF Regional Conf. Series in Appl. Math. Society for Industrial and Applied Mathematics, Philadelphia, PA, 1990.

[175] Y. Wang. Jump and sharp crusp detection by wavelets. *Biometrika*, 82:385–397, 1995.

[176] Y. Wang. Smoothing spline models with correlated random errors. *J. Amer. Statist. Assoc.*, 93(441):341–348, March 1998.

[177] N. Weyrich and G. T. Warhola. De-noising using wavelets and cross validation. In S.P. Singh, editor, *Approximation Theory, Wavelets and Applications*, volume 454 of *NATO ASI Series C*, pages 523–532, 1995.

[178] N. Weyrich and G.T. Warhola. Wavelet shrinkage and generalized cross validation for image denoising. *IEEE Transactions on Image Processing*, 7(1):82–90, January 1998.

[179] G. Winkler. *Image analysis, random fields and dynamic Monte Carlo methods.* Applications of Mathematics. Springer, 1995.

[180] Y. Xu, J. B. Weaver, D. M. Healy, and J. Lu. Wavelet transform domain filters: a spatially selective noise filtration technique. *IEEE Transactions on Image Processing*, 3(6):747–758, 1994.

Index

Software

Many figures in this volume were constructed with the aid of Matlab^{TM}. Nearly all of them can be reproduced using a collection of Matlab files, that can be downloaded from the following web site:

`www.cs.kuleuven.ac.be/~maarten/software/`

This software can be integrated into the software package WaveLab from Stanford University (see: `www-stat.stanford.edu/~wavelab/`), but it also stands on its own. This means: it provides its own routines for wavelet transforms in one and two dimensions, decimated and redundant, first and second generation, orthogonal and biorthogonai. The use of Matlab Toolboxes has been minimized. Special attention was paid to speed by using vector implementations (Matlab BLASS routines) whenever reasonable. An introductory demo, examples of test procedures and help comments are also included.

Other software packages include:

1. the already mentioned *Wavelab*, available free of charge from
   `www-stat.stanford.edu/~wavelab/`.
   Widely spread among statisticians, this collection of Matlab files basically covers most of the research at the Statistics department of Stanford University.

2. the *Rice Wavelet Tools*, see `www.dsp.rice.edu/software/`. They have Matlab code for filter bank wavelet design, for wavelet analysis, for wavelet domain Hidden Markov Random Fields and for wavelet based image deblurring/deconvolution. There is also Java code for image compression, using wavelets.

3. *WaveThresh* is an S-PLUS package for wavelet transforms in one and two dimensions and for thresholding. See
   `www.stats.bris.ac.uk/pub/Software.html`
   for more information.

4. *WAILI* (Piefpak) is a C++ library for two dimensional, integer wavelet transforms, including elementary image processing operations and thresholding using Generalized Cross Validation. The basic version can be downloaded free of charge from `www.cs.kuleuven.ac.be/~wavelets/`. An extended version aims at "very large images": these are images that do not fit into the working memory of a single computer. This version is available for research purposes only.

Lecture Notes in Statistics

For information about Volumes 1 to 108, please contact Springer-Verlag.

136: Gregory C. Reinsel, Raja P. Velu, Multivariate Reduced-Rank Regression. xiii, 272 pp., 1998.

137: V. Seshadri, The Inverse Gaussian Distribution: Statistical Theory and Applications. xii, 360 pp., 1998.

138: Peter Hellekalek and Gerhard Larcher (Editors), Random and Quasi-Random Point Sets. xi, 352 pp., 1998.

139: Roger B. Nelsen, An Introduction to Copulas. xi, 232 pp., 1999.

140: Constantine Gatsonis, Robert E. Kass, Bradley Carlin, Alicia Carriquiry, Andrew Gelman, Isabella Verdinelli, and Mike West (Editors), Case Studies in Bayesian Statistics, Volume IV. xvi, 456 pp., 1999.

141: Peter Müller and Brani Vidakovic (Editors), Bayesian Inference in Wavelet Based Models. xiii, 394 pp., 1999.

142: György Terdik, Bilinear Stochastic Models and Related Problems of Nonlinear Time Series Analysis: A Frequency Domain Approach. xi, 258 pp., 1999.

143: Russell Barton, Graphical Methods for the Design of Experiments. x, 208 pp., 1999.

144: L. Mark Berliner, Douglas Nychka, and Timothy Hoar (Editors), Case Studies in Statistics and the Atmospheric Sciences. x, 208 pp., 2000.

145: James H. Matis and Thomas R. Kiffe, Stochastic Population Models. viii, 220 pp., 2000.

146: Wim Schoutens, Stochastic Processes and Orthogonal Polynomials. xiv, 163 pp., 2000.

147: Jürgen Franke, Wolfgang Härdle, and Gerhard Stahl, Measuring Risk in Complex Stochastic Systems. xvi, 272 pp., 2000.

148: S.E. Ahmed and Nancy Reid, Empirical Bayes and Likelihood Inference. x, 200 pp., 2000.

149: D. Bosq, Linear Processes in Function Spaces: Theory and Applications. xv, 296 pp., 2000.

150: Tadeusz Caliński and Sanpei Kageyama, Block Designs: A Randomization Approach, Volume I: Analysis. ix, 313 pp., 2000.

151: Håkan Andersson and Tom Britton, Stochastic Epidemic Models and Their Statistical Analysis. ix, 152 pp., 2000.

152: David Ríos Insua and Fabrizio Ruggeri, Robust Bayesian Analysis. xiii, 435 pp., 2000.

153: Parimal Mukhopadhyay, Topics in Survey Sampling. x, 303 pp., 2000.

154: Regina Kaiser and Agustín Maravall, Measuring Business Cycles in Economic Time Series. vi, 190 pp., 2000.

155: Leon Willenborg and Ton de Waal, Elements of Statistical Disclosure Control. xvii, 289 pp., 2000.

156: Gordon Willmot and X. Sheldon Lin, Lundberg Approximations for Compound Distributions with Insurance Applications. xi, 272 pp., 2000.

157: Anne Boomsma, Marijtje A.J. van Duijn, and Tom A.B. Snijders (Editors), Essays on Item Response Theory. xv, 448 pp., 2000.

158: Dominique Ladiray and Benoît Quenneville, Seasonal Adjustment with the X-11 Method. xxii, 220 pp., 2001.

159: Marc Moore (Editor), Spatial Statistics: Methodological Aspects and Applications. xvi, 282 pp., 2001.

160: Tomasz Rychlik, Projecting Statistical Functionals. viii, 184 pp., 2001.

161: Maarten Jansen, Noise Reduction by Wavelet Thresholding. xxii, 224 pp., 2001.